The Reenchantment of Nature

The Reenchantment of Nature

The Denial of Religion
and the Ecological Crisis

Alister McGrath

Doubleday

New York   London   Toronto   Sydney   Auckland

PUBLISHED BY DOUBLEDAY
a division of Random House, Inc.
1540 Broadway, New York, New York 10036

DOUBLEDAY and the portrayal of an anchor with a dolphin are trademarks of
Doubleday, a division of Random House, Inc.

Book design by Patrice Sheridan

Library of Congress Cataloging-in-Publication Data
McGrath, Alister E., 1953–
The reenchantment of nature: the denial of religion and the ecological crisis /
Alister McGrath.—1st ed.
p.   cm.
Includes bibliographical references and index.
I. Human ecology—Religious aspects—Christianity.   I. Title.
BT695.5 .M444 2002
261.8'362—dc21   2002022256

ISBN 0-385-50059-9

PRINTED IN THE UNITED STATES OF AMERICA

September 2002
First Edition
10 9 8 7 6 5 4 3 2 1
BVG

Contents

chemistry at Oxford University, allowing me to interact with some of the finest scientific minds of the period. I went on to do research in molecular biophysics, trying to appreciate the complex physical processes that lay behind the workings of the human body and to develop new ways of understanding them. Contrary to what one might expect, my sense of wonder at the intricacies of nature was in no way diminished by my research into the physical and biological worlds. If anything, I found myself increasingly appreciative of the beauty and complexity of the world. It was easy to see why some people could dedicate a lifetime to the study of nature, and to sense something of the spiritual and intellectual intoxication that resulted.

So I set out on a voyage of discovery which still continues—a longing to make sense of the world in which we live and our own place within it. By studying the natural sciences, I was able to feed my relentless appetite to understand nature and appreciate its majesty. It seemed to me that the universe was something that teemed with significance and purpose. What greater privilege and excitement could there be than to engage with its wonders and mysteries?

Yet something was not quite right. Anxieties were already forming in my mind as I worked late into the night in Oxford's research laboratories. To investigate and appreciate the wonders of nature was one thing. But research in the natural sciences often seemed to be little more than a prelude to exploitation of nature, so that the *study* of nature all too often led directly to the *pillage* of nature. My colleagues in the laboratory in which I worked often quipped that the nice thing about "pure science" was that you never got your hands dirty. That was something best left to "applied science."

I could not simply ignore what was being done with the research that I and others were undertaking. What to me was an intellectually exhilarating adventure, born of a love of nature and a respect for its intrinsic complexity, was being corrupted into a means of exploiting that same nature for financial gain and the advancement of political goals. Despite this growing sense of moral unease, the deep sense of satisfaction I gained from research allowed me to overlook this disturbing side of the world of science. Some of those who taught me to love and value the natural sciences went on to paint the universe as purposeless and empty. Their emphasis on objective

Introduction

Nature has a rather splendid way of impressing us. There are few who have not been overwhelmed, time after time, by a sense of awe at the beauty of a glorious sunset, the sight of distant mountains shrouded in a soft blue haze of mist, or the brilliance of a starlit night. As a child, I found myself fascinated by the mysterious patterns of the constellations, the faint glow of the Milky Way, and the slow movements of the planets across the night sky. I devoured books on astronomy and even managed to build a small telescope to study the sky in greater detail. It allowed me to observe the moons of Jupiter, the lunar seas, and the phases of Venus.

Around the same time, I came into possession of an old microscope, which had belonged to a great-uncle who had once been head of pathology at the Royal Victorian Hospital, Belfast. I can still recall the spine-tingling sense of wonder when I first viewed the complex structure of the stamens of a daffodil I had just picked in the garden, and the intricate detail of the microscopic aquatic creatures I found in a local pond. It was as if a new world had been opened up to me, one that promised both spiritual and intellectual riches.

Unsurprisingly, I therefore chose to specialize in the mathematics and the natural sciences at high school. In 1971, I won a scholarship to study empirical knowledge certainly helped me know more about the universe. Somehow, though, it robbed nature of something immensely significant—the loss of which has dulled our sense of wonder at, and eroded our respect for, nature.

It is no accident that this erosion of a sense of wonder in the presence of nature coincides with the massive degradation and exploitation of nature through human agencies. In any relationship, a loss of respect is an essential preliminary to exploitation. The "disenchantment of nature" is one of the most worrying cultural developments of our time. What once evoked a sense of awe from appreciative and respectful human beings has now been explained away, deconstructed and desacralized. It is as if a protective veil has been torn aside from the face of nature, inviting exploitation of what now lies exposed in all its vulnerability.

This is a very serious matter, which has been met with various responses from concerned individuals. One highly influential approach is to insist that nature is sacred—to be thought of as a god or a goddess. Nature is therefore to be treated with the deference and respect appropriate to that divinity. This attitude can be found in the recent writings of the ecofeminist "Starhawk" (née Miriam Simos), a witch (the term she prefers) from California who urges her readers to see nature as a goddess.

> When I say "Goddess" I am not talking about a being somewhere outside of this world, nor am I proposing a new belief system. I am talking about choosing an attitude: choosing to take this living world, and creatures in it, as the ultimate meaning and purpose of life, to see the world, the earth and our lives as sacred.

In other words, we choose to see nature in these elevated terms, and revere and respect it accordingly. For Starhawk, this amounts to a decision on our part to see nature as divine. It is not that there is something intrinsically sacred about nature, which obliges us to view and interpret it in this manner. We choose to see nature as sacred, irrespective of whether nature itself demands this response from us.

A related approach to the ecological crisis essentially takes the form of an appeal to humanity's self-centered nature. If the human race is to survive, it will have to adopt a new and more respectful attitude to the natural world. That's not because there is anything special about nature. It's just that a failure to respect it will lead to an environmental catastrophe that will probably wipe us all out. If, with Starhawk, we choose to see the world as sacred, we are likely to treat it in a more respectful way which will indirectly safeguard our future. So it makes sense to treat nature with respect. The survival of humanity is, after all, a pretty compelling argument. The credibility of this approach to nature lies more in its consequences than in its foundations. Yet there is no doubt that many find it attractive.

But what if there was something about nature itself that calls out for us to treat it in this way? What if the deep structure of nature was such that we were forced to realize that it *is* special? The deferential treatment of nature would then rest not just on the consequences of this belief—as if this was a matter of convenience for us—but on account of the deep structure of reality itself. There is a third approach, which argues that a right attitude to nature rests on the revival of our capacity for wonder, resting on our appreciation of the nature of reality itself. If nature has become *disenchanted*, the remedy lies in its *reenchantment*. We do not need to invent some distinctive quality about nature, nor do we need to impose an alien framework upon nature to see it as sacred. It already is special—but awaits our appreciation of this fact. The challenge lies with us. Are we willing to recognize the hidden depths of nature and their implications for us?

This book is about encountering nature in all its fullness, and the readjustments to our settled ways of thinking that this demands. We live in a world that is dismissive and intolerant of the view that nature is something to be loved rather than just studied ("sentimentalism"), or the idea that nature can point to a deeper level of existence and meaning ("superstition"). So how might this come about?

I found my attitude toward nature changing profoundly during my student days at Oxford. As a high school student in Ireland, I had come to regard religion as little more than outmoded superstition. My initial view had been that it was harmless stuff, which shielded inadequate people from the harshness of life. Like many young people growing up in the 1960s, however, I began to become intrigued with Marxism. I came to see religion in stridently negative terms, as an evil force which held back intellectual and social progress. The student riots of 1968 seemed to signal that a new era in Western history was about to dawn, in which outmoded habits of thought—such as religion—would be replaced by more progressive ways of thinking and living. I actually knew virtually nothing about religion, as it happened. But that didn't seem especially important at the time. My Darwinian instincts told me that atheism would survive and finally triumph. And who needs to *understand* losers? That seemed to sort things out admirably. It proved exceptionally easy to discard something that I did not really know or understand. Yet a slight sense of unease remained with me. Had I dismissed religion too quickly?

When I arrived at Oxford, I decided to sort myself out intellectually, once and for all. I began to explore the ideas and appeal of Christianity in much more detail, assuming that this would allow me to confirm my vigorous rejection of its intellectual and spiritual viability. Unfortunately, things did not work out quite as I had anticipated. It was easy to dismiss Christianity without knowing it particularly well, or after judging it on the basis of its rural Irish forms. Encountering it at Oxford University, in an intellectually robust and spiritually enriching form, caused me to rethink things seriously. Oxford atheism, in contrast, seemed plodding and dull, with student interest focusing on the sexual habits rather than the leading philosophical ideas of one of its chief practitioners, A. J. Ayer.

In the autumn of 1971, I followed in the footsteps of my fellow countryman, C. S. Lewis, and gave in to what I regarded as the coherence and attraction of the Christian faith. This unexpected and not entirely welcome development completely threw me. I found that my rather comfortable and settled mental world was shaken, forcing me to wrestle with new questions and reopening questions that I thought I had settled long ago. It was an immensely exciting and rewarding time, in which I began to explore new perspectives on life and nature.

In particular, my newfound Christian faith brought a new sense of fulfillment and appreciation to my studies in the natural sciences. While some argue that religion alienates people from science, I found these two areas of human thought converging to yield a dynamic flux that was at once immensely spiritually satisfying and intellectually exciting. My attitude to the natural world changed as a result, with my engagement with nature becoming more intense and committed. I now knew that nature was charged with the grandeur and majesty of God. To engage with nature was to gain a deeper appreciation of the divine wisdom. Any good natural scientist will speak of a sense of wonder arising from the study of the depths of nature; I found that my existing feelings of awe were intensified and given a new sense of purpose by the Christian faith. Influenced by Christian writers such as C. S. Lewis, I began to develop a strong sense of nature as something that pointed beyond itself—a sign of something mysterious beyond it, which seemed to bring nature's potential to fulfillment.

The most profound change I noticed was my attitude toward the integrity of nature. I discovered that humanity's belief in its inalienable right to exploit nature was negated by the fundamental Christian idea of creation. If nature was God's, humanity was in no position to mess around with it. Humanity's role was that of a steward or caretaker of creation. Far from being owners or masters of the created order, and free to behave as we pleased, we were charged with the tending of creation and were accountable for our actions. This heightened my growing unease over the uses to which the natural sciences were put by humanity, and convinced me of the need to work toward environmental responsibility and accountability.

Where some natural scientists choose to remain active in their fields of research, exploring issues of spirituality or theology from within that competency, I took the decision to cease active scientific research in order to develop an expertise in Christianity. I moved to Cambridge University to undertake research in the relation between science and Christianity. My initial idea had been to explore this interaction throughout history, but I moved away from this when it became clear that I would need to master the history of Christian thought—the field traditionally known as historical theology—before I would be able to make any reliable contribution to the field of science and religion in general, and ecological concerns in particular. By 1996, after nearly twenty years of research in historical theology, I felt that I had achieved such a competency.

I now began to focus on environmental issues, wishing to bring the rich and resilient resources of the Christian tradition to bear on this immensely important problem. To my astonishment, this proved a frustrating experience. Some environmentalists muttered slogans like "Christianity is part of the problem!" as if they were some kind of sacred mantra. Christianity and the churches were either written off as an irrelevance or excoriated as actually *encouraging* the rape of the earth. I found this astonishing. Environmentalism needs all the friends it can get, and it makes little or no sense to write off potential supporters with this kind of nonsense. But much more seriously, it brought home to me how misinformed many within the ecological movement were about the nature of the Christian faith and its positive implications for ecological reflection and action. Having taken some trouble to master the Christian tradition, I was deeply troubled by the levels of prejudice, ignorance, and misunderstanding I encountered.

It did not take long to work out what the problem was. The intellectual roots of this attitude turned out to be surprisingly shallow, and actually rested upon one single work, which seemed to have achieved cult status within the environmental movement. In 1967, a historian at the University of California published a short article that would shape the attitudes of a generation. In "The Historical Roots of our Ecological Crisis," Lynn White seemed to lay the blame for the emerging ecological crisis firmly and squarely on Christianity. The sound bite that his many readers took away with them? "We shall continue to have a worsening ecological crisis until we reject the Christian axiom that nature has no reason for existence save to serve man."

These were bold and simple words, written at a formative stage in the emergence of the modern environmental movement. Attitudes were being shaped, and beliefs molded. There were few, if any, older voices to mentor a new generation of environmentally concerned individuals. In the absence of wise mentors, a handful of core texts served as their guides. White's article opened and closed their discussion of the role of religion in ecological issues. A scapegoat had to be found for the ecological crisis, and this article conveniently provided one. Where there is a problem, there is a perpetrator. It's just a question of naming, blaming, and shaming the guilty party.

Christians were to blame for the environmental crisis. As many people within the movement had no time for Christianity anyway, White's article suited their purposes well. It crystallized an outrageous perception, which persists to this day: Christianity is the enemy of the environment.

So yet another myth, such as Jews being responsible for the economic problems of 1930s Germany, was added to the long list of destructive influences on our culture. Western religion in general, and Christianity in particular, came to be demonized as the perpetrator of our emerging environmental crisis. The Judeo-Christian tradition was branded as the cause of the ecologically disastrous transition from a respectful attitude toward nature to the Western technological revolution, with its devaluation of the natural order and legitimation of human dominance. It's hard to exorcise that demon when it has become such an integral part of so many people's core beliefs. I have spoken to many environmentalists to emerge from this halcyon period, and without exception they identify White's article as shaping their often highly negative views on the impact of religion on ecology. It became a sacred text, passed on with an almost uncritical reverence. Many ecologists would no more criticize White's article than fundamentalist Christians would criticize the Bible.

Which is a pity. There were unquestionably some genuine insights in that article, to be valued and cherished. One of these stands out as being of supreme importance. With deadly accuracy, White pinpoints the importance of religion in relation to ecology: "What people do about their ecology depends upon what they think about themselves in relation to things around them. Human ecology is deeply conditioned by belief about our nature and destiny—that is, by religion." Yet White's recognition of the critical importance of religion to ecology was interwoven with what seemed like a hasty and uncritical condemnation of Christianity, resting on a highly questionable and selective reading of the Christian tradition. My own intensive research on the Christian understanding of nature gave me no reason whatsoever to suppose that the complex Christian tradition could be summed up in the simplistic assertion that "nature has no reason for existence save to serve man." I was saddened that such stridently negative views about the role of my faith community had been injected into such an important discussion. Saving the earth is so important that there is simply no place for polemicizing against potential allies in this struggle—especially when this critique is so inaccurate.

That's why I decided to write this book. It is intended to bring out the strategic resources of the Christian faith for the environmental struggle, mindful of the diversity of Christianity and the distinctive contributions of its various components. It is written from within a community of faith, in which the basic themes of the Christian faith are taken seriously and allowed to impact both life and thought. I hope to challenge Christians to take ecological issues more seriously and to help those outside the Christian faith to appreciate its important contribution to these issues.

While more than five out of six Americans identify with Christianity, some environmentalists have embarked on a search into Eastern and Native American beliefs to find a solution to our ecological predicament. Having read—or, in some cases, I fear, merely read about—Lynn White's article, many "deep ecologists" believe that Christianity is the enemy of environmentalism. This totally spurious conclusion rests on a failure to subject White's argument to the critical examination it demanded. Instead, they have turned to Eastern religions—or, more accurately, the highly eclectic, domesticated, and sanitized versions of these religions, which often bear little resemblance to their Asian forebears, especially in relation to their evaluation of nature.

Doubtless there is something to be gained from this. But might not the careful study of the Christian tradition offer something that might carry greater weight for most Americans? There are unquestionably points at which Christians need to be critiqued and challenged in these matters, and I shall not hold back from doing so. Where White makes valid criticisms of Christianity, these need to be heard respectfully and given careful examination.

The basic idea that this book sets out to explore is that the grounds of our ecological crisis lie in the emergence of a worldview that proclaimed human autonomy and viewed nature as a mechanism subordinated to humanity. Christianity may well be implicated in the emergence of this worldview, as we shall see. Yet once it had emerged, this worldview declared its independence of any intellectual and moral constraints and set out to go its own way. Shunning its intellectual forebears as something of an intellectual embarrassment, it turned its back on the classic Christian notion of the limitation of human engagement with nature and the accountability of humanity for the outcome of such an engagement. The result was the elimination of a fundamental sense of accountability for the human use of nature and the rejection of any idea of nature being special as "outmoded superstition." Nature, once seen as having a privileged status, was *disenchanted*. To understand how nature can be *reenchanted*, we must first appreciate how it came to be robbed of its special status. We must therefore ensure that our exploration of our present crisis probes the emergence of this worldview, and its very real implications for our situation.

This book, then, explores why religion is essential if we are to fully grasp, appreciate, and respect nature. It aims to persuade Christians that they ought to be taking nature a lot more seriously, and anyone concerned with nature that they ought to be taking Christianity a lot more seriously than they have to date. But above all, this book aims to set out the intellectual excitement of engaging with nature and recovering that lost sense of wonder. Perhaps it is not lost, after all. Perhaps it is just buried under some debris, waiting to be rediscovered and renewed.

Alister McGrath
Oxford, January 2002

The Reenchantment of Nature

CHAPTER ONE

The Meaning of Life and Other Enigmas

What is life all about? Does it possess any intrinsic meaning? Or is this "meaning" just something we impose upon a meaningless void? These are sincere and important questions, and there has been no shortage of answers. In his comic masterpiece *Hitchhiker's Guide to the Galaxy*, Douglas Adams tells how a race of superintelligent beings from a very advanced civilization constructed a supercomputer called Deep Thought to answer the question "What is the meaning of life, the universe, and everything?" Deep Thought's circuit boards pulsed with activity for seven and a half million years and finally produced a result: 42—an enigmatic answer, to say the very least. Perhaps an even more puzzling answer is offered in Alan Dean Foster's *Glory Lane*, which tells of a group of people who visit another advanced civilization and ask its similarly advanced supercomputer more or less the same question. This time, the meaning of life is defined in less numerical, but still slightly baffling, terms: shopping.

Perhaps these answers are meant to caution us concerning the reliability of some of the more serious answers to this question. Precisely because these answers are of such importance, people tend to treat them with suspicion, even cynicism. And they are right to do so. How many people have been deluded, hoodwinked, or pressured into accepting less than adequate, and even dangerous, answers? Yet this understandable degree of cynicism must not force us to draw the conclusion that there is no meaning to life; or that, if there is indeed a meaning, it is so hidden and obscure that none can hope to find it.

Many of the answers given to these questions are religious, and for that reason they automatically attract ridicule from the "let's get rid of religion" school. Oxford zoologist Richard Dawkins, who is a particularly luminous representative of this group, is quite clear why so many people find religion attractive. It offers them—to use his terms—"explanation," "consolation," and "uplift." In every case, of course, Dawkins argues that what religion offers is completely false and that the truthfulness of the sciences is to be preferred. Science may not always be able to offer the equal of religion, but at least what it offers is absolutely true. As Dawkins puts these points in an article in *Humanist* magazine, following his election as Humanist of the Year:

> Humans have a great hunger for explanation. It may be one of the main reasons why humanity so universally has religion, since religions do aspire to provide explanations. We come to our individual consciousness in a mysterious universe and long to understand it. Most religions offer . . . a cosmology and a biology; however, in both cases it is false.
>
> Consolation is harder for science to provide. Unlike religion, science cannot offer the bereaved a glorious reunion with their loved ones in the hereafter . . .
>
> Uplift, however, is where science really comes into its own. All the great religions have a place for awe, for ecstatic transport at the wonder and beauty of creation. And it's exactly this feeling of spine-shivering, breath-catching awe—almost worship—this flooding of the chest with ecstatic wonder, that modern science can provide.

One cannot help but feel that Dawkins here proceeds from preconceived ideas to predetermined conclusions with almost indecent haste. Religion is "false"—just like that, totally and completely. No room for doubt, discussion, or debate. The possibility of even a hint of the truth in any religion is excluded as a matter of principle. Nothing that religion says can *possibly* be right. Dawkins seems to live in a chiaroscuro world, in which everything is black and white, true or false. Science is true, religion is false—a neat little slogan, with about as much plausibility as George Orwell's creed from *Animal Farm:* "two legs bad, four legs good."

One of the reasons that religion is enjoying new public interest is a new awareness in its ability to address questions about what gives life its meaning, what consoles us in moments of darkness and despair, and what gives us a sense of hope, vision, and wonder. It is almost as if we are preprogrammed to ask these questions and deem them important. As the philosopher Bertrand Russell famously remarked in his *History of Western Philosophy,* once humanity has managed to work out how to feed itself, it naturally turns its attention to the great questions of meaning and significance. How do we fit into the greater scheme of things? These questions underlie the new interest in spirituality that has swept through much of the Western world in recent years.

The Recovery of Religion

One of the most distinctive features of the last few decades has been a rediscovery of the spiritual dimension to human existence. In every continent of the world—with the conspicuous exception of western Europe—there has been a surge of interest in the concept of the transcendent, with a growing reaction against what are seen as the unsatisfactory reductionisms of various materialist philosophies and worldviews. Why has there been this cascade in interest in spirituality and religion? Social theorists offer us explanations that sometimes sound rather like academic sour grapes— religion is just what infantile minds hang on to in times of crisis, or what people turn to in an attempt to resist modernization. Sure. Yet there are other more compelling reasons, which help us understand why religion will continue to play such a major role in human life and culture.

Religion offers explanation, consolation, and inspiration in about equal measure. Each of these resonates with a fundamental aspect of human life and thought. For the Christian, this is entirely predictable. If the world and humanity are created by God, such resonance is to be expected, as we shall see in a later chapter. Everyone wants to find something that is really worth pursuing and possessing rather than the chimeras whose luster vanishes once they are secured. The greatest questions life has to offer can be summed up in just a few words. What is *really* worth possessing? And where is it to be found?

These questions dominate the "wisdom literature" of the ancient Near East and far beyond. In one of the Gospel parables, Jesus compared the kingdom of heaven to a pearl of great price. "The kingdom of heaven is like a merchant looking for fine pearls. When he found one of great value, he went away and sold everything he had and bought it" (Matthew 13:45–46). On finding a beautiful and precious pearl for sale, the merchant realizes that he *must* sell everything else in order to possess it. Why? Because here is something of supreme value, something really worth possessing. Everything else seems of little value in comparison.

The merchant searching for that pearl is himself a parable of the long human quest for meaning and significance. It is clear from the parable that he already possesses many small pearls. Perhaps he bought them in the hope that they would provide him with the satisfaction that he longed for. Yet what he had thought would satisfy him proved only to disclose his dissatisfaction and make him long for something that was, for the moment, beyond his grasp. Just as the brilliance of the sun drowns that of the stars, so that their faint light can only be seen at night, so this great pearl allowed the merchant to see what he already owned in a different perspective.

For C. S. Lewis, the discovery of Christianity was like taking hold of and possessing something intrinsically precious and beautiful, which allowed the rest of the world to be seen in its reflected radiance. He put the significance of his discovery like this: "I believe in Christianity as I believe that the sun has risen—not only because I see it, but because by it I see everything else." Christianity offers a spine-tingling vision of the transcendent and a framework that helps make sense of life's joys, cruelties, ironies, and pain. For Lewis, Christianity is more than a theory in which one can take intellectual delight, offering a new appreciation of the beauty of the world—to be compared to Newton's optics or laws of motion or Maxwell's electrodynamic equations. It points to something that transcends these, which can be intuitively grasped in the present and which will be fully possessed in the future. The beauty of the world is affirmed, and declared to be a foreshadowing of the greater glory to come. As the great English religious poet George Herbert (1593–1633) put it, we are enabled to catch a glimpse of "heaven in ordinary."

Lewis attempted to put this into words by using an image from his childhood, when building sand castles on a trip to the seaside was seen as heaven itself:

> Our Lord finds our desires not too strong, but too weak. We are half-hearted creatures, fooling about with drink and sex and ambition when infinite glory is offered us, like an ignorant child who wants to go on making mud pies in a slum because he cannot imagine what is meant by a holiday at the sea. We are far too easily pleased.

For Lewis, the sense of wonder we experience at nature is not meant to satisfy us; it is meant to make us yearn for the greater wonder that it silently signposts and whispers will one day be ours.

All this seemed a little unrealistic in the light of the wave of secularism that swept through Western culture after Lewis's death in 1963. Secularizing social theorists predicted the coming of a future secular global culture with much the same confidence as an earlier generation of Soviet theorists proclaimed the historical inevitability of Marxism-Leninism. Religion was on its way out. Yet as William S. Bainbridge and Rodney Stark point out in their excellent critical study *The Future of Religion: Secularization, Revival, and Cult Formation*, "the most illustrious figures in sociology, anthropology and psychology have unanimously expressed confidence that their children—or surely their grandchildren—would live to see the dawn of a new era in which, to paraphrase Freud, the infantile illusions of religion would be outgrown." The inability of Marxism—or anything else, for that matter—to wipe out religion, by force of argument or force of arms, was one of the most significant landmarks of the twentieth century. The simplistic predictions of 1960s ivory-tower sociologists that the twentieth century would end in a tidal wave of secularism surging relentlessly around the globe has been shown to be hopelessly wide of the mark.

The new interest in spirituality in the West—which is not limited to traditional religious beliefs or practices—has led to a growing impatience with what is seen as the reductionism of some natural scientists, who improperly move from science as the investigation of reality to science as the determinant of what reality is in the first place. One can have nothing but respect for the natural scientist who urges us to understand, appreciate, and explore the intricacies of the natural order. But when some—let us be clear, a tiny yet loud minority—insist that what the sciences uncover is *all* that there is to life, we have every right to insist that they have strayed out of their field of competency. The growing interest in religion has had a significant impact on the popular estimation of the sciences and their capacity to serve humanity.

Perhaps one of the more interesting indicators of this shift is the change in direction of the long-running television series *Star Trek* and its wide-screen spin-offs. Classic *Trek* episodes from around 1966–69 were strongly influenced by the humanist philosophy of their creator, Gene Roddenberry. Roddenberry was basically an old-fashioned rationalist, who would have got on well with Thomas Jefferson and his circle of *intimes*. These early *Trek* episodes were notable for their uncritical affirmation of the excellence of science, the triumph of logic, and the inevitability of progress. One can easily imagine Richard Dawkins lending a little intellectual muscle to the crew of the *Enterprise*, not least in their fearless criticism of the greatest evil that the galaxy faced—religion. It was just a matter of time, these early episodes hinted, before the great cosmic scourges of religion, disease, and poverty were all eliminated and the galaxy would exude sweetness and light at every point.

The *Enterprise* was thus the flagship of scientific rationalism, adrift in a mad universe whose inhabitants persisted in believing the strangest things. In the heyday of 1960s rationalism, religion was viewed as one of the evils—along with poverty, prejudice, and war—that progress would leave behind. Religious beliefs were to be expected among the primitive alien societies favored by a visit from the crew of the starship *Enterprise*. But there could be no question of these enlightened and thoroughly modern progressives themselves having religious beliefs. Religion was best left to the savages of the most backward parts of the galaxy. While America rediscovered religion in the 1970s and 1980s, the *Enterprise* was still busy spreading its 1960s "no religion" message. This may have gone down well with the Klingons, but it was looking increasingly outdated back on planet earth.

It was not until Roddenberry's death in 1991 that this somewhat quirky outdated rationalism lost its hold over the program. The series now aligned itself with what was happening in American culture of the 1990s rather than the time warp of the 1960s. Science and progress were toppled from their throne, as a new interest in spirituality flourished within America. While *Trek* deliberately avoided the question of what religious beliefs were true, the message was clear: spirituality was a good thing, which cultured human beings needed, and should not hastily discard. Where Dr. Spock relied upon logic, Commander Chakotay of the USS *Voyager* preferred to trust in spirit guides.

This new interest in things spiritual has swept through all aspects of Western culture—not just TV—in the past decade. The burgeoning bookstore sections dealing with "Body, Mind, and Spirit" are a telling indicator of a shift in Western thought away from the world of the Enlightenment. This has not exactly been welcomed by old-fashioned rationalists, who have seen their cherished deities of reason and logic dethroned, to be replaced with angels, spirits, spiritual forces—not to mention Christianity.

This can perhaps be lazily dismissed as simply an irrational phase in Western culture, which will give way to something more coherent and realistic in the longer term. But it is not quite as simple as that. There may well be a significant degree of confusion within the new concern for spirituality that has swept through Western culture. Yet at its heart, this trend reflects a growing disenchantment with the overstatements of some natural scientists, which have increasingly caused disquiet within the culture at large.

One might be tempted to write off this growing concern as crude and uninformed antiscientific polemic. It is not. It is a deadly serious expression of concern over the way the natural sciences are being used, often with the collusion of some scientists who ought to know better. The natural sciences are to be respected and honored for their commitment to advancing knowledge, understanding, and insight concerning the structures of the universe in which we live. I continue to regard my time spent as an active research scientist to be among the most fulfilling and exciting periods of my life. The problems begin when the sciences—or *any* discipline, for that matter—start to see themselves as the unacknowledged rulers of intellectual empires, and others (especially those who presume to criticize them) as social and intellectual inferiors who still live back in the Stone Age.

One of the chief reasons for this growing skepticism concerning the value of the sciences is the intolerance and stridency of the antireligious polemic of some natural scientists, whose excoriations of the religious beliefs of the vast bulk of humanity have caused many to draw negative conclusions concerning the place of the natural sciences in the greater scheme of things. Antireligious scientists seem to slip with alarming ease from evidence-based theories to dogma-based worldviews, implying that the latter are as intellectually robust as the former. The simple fact of the matter is that the natural sciences neither *necessitate* nor *preclude* religious belief. To insist that they do either is to go far beyond the evidence available and impose a dogmatic worldview on an essentially multivalent reality, capable of more than one interpretation. So important is this point that we may explore it further.

Longing for an Explanation

One of the most fundamental of human impulses is a desire to explain. Indeed, one of the shared motivations of religion and the natural sciences is a longing to make sense of things. In his important book *Why Religion Matters* (2001), religion scholar Huston Smith tells of an encounter with the noted physicist David Bohm, responsible for some significant advances in quantum theory. Their argument was over whether there were limits to science. Huston took the view that an influential misreading of science "belittles art, religion, love and the bulk of the life we directly live by denying that those elements yield insights that are needed to complement what science tells us." Bohm argued that science had no limits, which prompted Huston to ask what Bohm meant when using the word "science." Bohm's response was instructive: "opennness to evidence." Huston riposted that this made him a scientist as well, a possibility that Bohm was apparently prepared to concede.

There is an important lesson to be learned from this conversation. Since the dawn of history, humanity has tried to make sense of the evidence around it—the movements of the stars and planets, the patterns of order within the world, and the mysteries of human behavior. The great wisdom literature of the ancient Near East can be seen as an attempt to discern the patterns lying beneath the occasionally baffling experiences of life. The explanations offered vary; the commitment to offer such answers is universal. Sadly, Richard Dawkins merely recycles the tired and dated myths of antireligious positivism, insisting that science alone is evidence-based and proves its beliefs, whereas religion ignores the evidence and is based on blind faith. The reality is quite different. As Huston Smith puts it, the world is ambiguous; it is like an inkblot, which can be perceived in different ways. Or, to put it in the rather more sophisticated terminology of philosopher Roy Bhaskar, reality is "multi-layered" or "stratified," with each layer demanding to be investigated and represented in different ways. An explanation that works well at one level doesn't necessarily work at another level.

Some of the explanations of life on offer are just too neat to be true.

They seem to ignore the messiness of life. While many people are attracted to simple solutions to complex problems, others view these with suspicion. We have become cynical about solutions that are just too neat to be credible and that claim to explain everything. We are as weary as we are wary of too-confident answers to difficult questions. Freudianism and Marxism are excellent examples of ambitious theories that can account for just about everything that happens. The situation is neatly illustrated by the often-told story of the man who visited his Freudian analyst. If he was early, the analyst concluded that he was anxious; if he was on time, he was compulsive; if he was late, he was resentful. The theory accounted for *everything*. It was precisely this inability to be refuted by experience that prompted Karl Popper (1902–94) to develop his celebrated theory of falsification. Irrefutability and an ability to explain everything might seem to be a virtue; in reality, it is a vice.

The world we experience is just too messy and fuzzy to fit completely into the orderly systems that some crave and others fear. Some find this idea infuriating and demand conformity to their own perspective on reality, even when this is clearly biased and skewed—for example, the materialist reductionisms that some pass off as "scientific" worldviews. How can we live when we cannot be confident of anything? The only certainty of our age seems to be that there is no certainty at all. Constant change is here to stay. Yet even this confident assertion contradicts itself, just like the statement that Bertrand Russell recalled seeing written on a college blackboard: "All statements written on this blackboard are false."

In the end, we just have to learn to live with an untidy world in which we cannot be certain of everything—a world in which there are unanswered questions. As Huston Smith points out, "people have never agreed on the world's meaning, and (it seems safe to say) never will." We are faced with "life's cosmic inkblot," which can be interpreted in many ways and at many levels.

One of those ways is to see the world as God's creation and use this as a framework for making sense of the natural world we inhabit and our own place within it. In what follows, we shall explore how the Christian doctrine of creation has a direct bearing on our understanding and appreciation of nature. As we shall see, the doctrine of creation is pregnant with ecological insights and brings intellectual enrichment to the interaction with the natural world.

A Christian Doctrine of Creation

Christianity offers an explanation of the world and our place within it. In common with every other view of the world, it is not a viewpoint that commands universal assent. One of the central Christian perspectives on the world concerns the doctrine of creation. What difference does this make to the way the world is perceived, experienced, and valued?

The Old Testament uses a number of illustrations to illuminate the idea of creation. The rich imagery of the Genesis creation accounts is supplemented by many other biblical passages that portray God as a master builder, deliberately constructing the world (for example, Psalm 127:1). The imagery is powerful, conveying the ideas of purpose, planning, and a deliberate intention to create. The image allows the beauty and ordering of the resulting creation to be appreciated, both for what it is in itself and for its testimony to the creativity and care of its creator. Nature can thus be viewed or "seen as" (N. R. Hanson) the work of a master craftsman, and the wisdom of the creator can thus be seen in the ordering of the world, in much the same way as the wisdom of an architect can be seen in the design of a great building. Creation is thus about imposing structure upon reality—a structure that the human mind can subsequently perceive and grasp.

One of the most significant parallels between the natural sciences and Christianity is this fundamental conviction that the world is characterized by regularity and intelligibility. The Christian faith holds that the patterning and regularity of the world are not imposed upon that world by an order-seeking mind, but are inherent to nature itself. As the leading Australian physicist Paul Davies points out in *The Mind of God,* "in Renaissance Europe, the justification for what we today call the scientific approach to inquiry was the belief in a rational God whose created order could be discerned from a careful study of nature."

Nature thus bears witness to God's wisdom, just as a great building bears witness to the genius of its designer. St. Paul's Cathedral, London, is one of the greatest works of the architect Sir Christopher Wren (1631–1723). The cathedral had to be rebuilt after the Great Fire of London (1666), and the task of designing the new building was entrusted to Wren. The great new building was finally completed in 1710 and remains one of the most famous landmarks of London. There is no memorial to Wren in that cathedral. In its place, there is an inscription over its north door: "If you are looking for a memorial, look around you." The genius and wisdom of the architect can be seen in what he built.

In the same way, the wisdom of God can be discerned within the creation, which is a witness to the power and wisdom of its creator. In the words of the Psalmist, "The heavens are telling the glory of God" (Psalm 19:1). As the French Renaissance thinker Jean Bodin (1539–96) put it in his *Universae naturae theatrum* (The Theater of the Universe of Nature):

> We have come into this theater of the world for no other reason than to understand the admirable power, goodness and wisdom of the most excellent creator of all things, to the extent that this is possible, by contemplating the appearance of the universe and all his actions and individual works, and thus to be swept away more ardently in praise of him.

A similar point was made centuries later in an article in *Nature* by J. J. Thomson, the discoverer of the electron: "In the distance tower still higher [scientific] peaks which will yield to those who ascend them still wider prospects and deepen the feeling whose truth is emphasized by every advance in science, that great are the works of the Lord."

A second way of thinking about divine creation is to compare it with the creative actions of an artist—like someone painting a picture or composing a symphony. The image of God as artist conveys the idea of personal expression in the creation of something beautiful. We might speak of such an artist "putting a lot of herself into" a picture or music, meaning that the artistic creation in some way mirrors the nature and genius of the artist. Yet the same God who created the universe also created us. There is thus a created resonance between ourselves and the universe. We are enabled to hear the music of its creator and discern the hand of the creator within its beauty. It is part of the purpose of the creator that we should hear the music of the cosmos and, through loving its harmonies, come to love their composer.

The American theologian Jonathan Edwards (1703–58) developed this theme at several points in his writings. In his *Personal Narrative,* Edwards wrote of his "sheer beholding of God's beauty" as he walked in the New Jersey countryside:

> As I was walking there and looking up into the sky and clouds, there came into my mind so sweet a sense of the glorious *majesty* and *grace* of God, that I know not how to express. I seemed to see them both in a sweet conjunction . . . it was a sweet and gentle and holy majesty; and also a majestic meekness.

Once Edwards learned to see nature as God's creation, he found the glory of God being reflected back to him in even the most modest wonders of nature. It was as if the natural world tempered God's glory, enabling the human eye to behold what would otherwise be inaccessible to our weak gaze. Edwards writes thus of the transformation in perceptions resulting from seeing nature as God's creation:

> The appearance of every thing was altered; there seemed to be, as it were, a calm sweet cast, or appearance of *divine glory, in almost every thing.* God's excellency, his wisdom, his purity and love, seemed to appear in every thing; in the sun, moon, and stars; in the clouds, and blue sky; in the grass, flowers, trees; in the water, and all nature; which used greatly to fix my mind. I often used to sit and view the moon for continuance; and in the day, spent much time in viewing the clouds and sky, to behold *the sweet glory of God* in these things; in the meantime, singing forth with a low voice my contemplations of the Creator and Redeemer.

The beauty of nature thus became a pointer to the glory of God. This sense of aesthetic ecstasy pervades Edwards's more autobiographical writings, such as his *Miscellanies*. The perception of beauty that we experience "when we are delighted with flowery meadows and gentle breezes" is, for Edwards, nothing other than an intimation of the beauty of God.

This does not mean that everyone, looking at the cosmos, draws the conclusion that it is the creation of God, reflecting something of God's character and wisdom. In one sense, the world is like a parable. Parables, like those told by Jesus, can be understood at two levels. Although the old definition of a parable as "an earthly story with a heavenly meaning" is a little simplistic, it serves to make an important point—that the spiritual meaning can be missed. The parable of the shepherd, for instance, who goes out in search of a lost sheep and brings it home on his shoulders (Luke 15:3–5) can be read as a touching story based on the Palestinian rural economy of the first century. But to those in the know—or, to use the turn of phrase we find in the Gospels themselves, "to those to whom the secret of the kingdom of God has been given" (Mark 4:11)—the story tells of the love of a God for a wounded and wayward humanity, which compelled God to go in search of us and bring us safely back to the place where we belong.

The same point, of course, can be expressed in different terms. William Golding's novel *Lord of the Flies* (1954) can be read in many ways—as a gripping narrative of a group of boys on a desert island or as "an attempt to trace the defects of society back to the defects of human nature" (Golding). It is perfectly possible to read the novel at one level and miss the other entirely. The deeper meaning is something that is to be discovered; once it has been grasped, the novel can never be read in the more superficial way again. The deeper meaning acts like a lens or pair of spectacles, leading us to see it in a certain way which others, reading the same novel, might miss.

The same point applies to the world. It can be viewed at one level as a material entity, whose origins and behavior can be accounted for by the laws of physics. But this most emphatically does not rule out the discernment of a deeper level. The great scientific writers of the Renaissance—such as

Kepler and Galileo—spoke of the natural world as being a "book." For Galileo, this book was written in the universal language of mathematics:

> Philosophy is written in this grand book, the universe, which stands continually open to our gaze. But the book cannot be understood unless one first learns to comprehend the language and read the letters in which it is composed. It is written in the language of mathematics.

This book, like *Lord of the Flies*, can be read at different levels. Some will see it as God's creation, others not. The twelfth-century Parisian writer Hugh of St. Victor stated this point as follows:

> For the whole physical world is a kind of book written by the finger of God—that is, created by divine power. Each particular creature is like an image—not an image invented by some human decision, but instituted by the divine will to reveal the invisible things of God's wisdom. And just as an illiterate person, on seeing an open book, would see its images but not be able to make any sense of its words ... so some may see the outward appearance of these visible creatures, without understanding their inward meaning.

So what perspectives does the Christian bring to the study and care of nature? Perhaps one of the most important of them concerns the place of humanity, and above all its capacity to understand nature and respond to its beauty. We shall explore this in what follows, focusing especially on the human sense of wonder at the beauty of nature.

Reflections of God in Nature

One important insight that Christians derive from a doctrine of creation, as we saw above, is that nature possesses an inbuilt ability to act as a sign to its creator. The divine creation of the world establishes an analogy between the creator and what is created. The beauty of the world thus reflects the beauty of God. Nature is like a mirror, itself beautiful while reflecting an even greater beauty of God. To study the wonder of nature is to glimpse tantalizing facets of the face of God, and long to see more. As the great medieval theologian Thomas Aquinas (1225–74) put it:

> Meditation on God's works enables us, at least to some extent, to admire and reflect on God's wisdom... We are thus able to infer God's wisdom from reflection upon God's works... This consideration of God's works leads to an admiration of God's sublime power, and consequently inspires reverence for God in human hearts... This consideration also incites human souls to the love of God's goodness... If the goodness, beauty and wonder of creatures are so delightful to the human mind, the fountainhead of God's own goodness (compared with the trickles of goodness found in creatures) will draw excited human minds entirely to itself.

Something of the torrent of God's beauty can thus be known in the rivulets of the beauty of the creation. This has long been recognized as one of the most basic religious motivations for scientific research—the passionate belief that to gain an enhanced appreciation of the beauty of the world is to glimpse something of the glory of God.

The importance of the quest for beauty in the natural sciences has been stressed by many distinguished physicists and mathematicians. One of the most powerful affirmations of the role of beauty in science, particularly as a motivating factor, is offered by S. Chandrasekhar of the University of Chicago, who won the Nobel Prize in physics. In his 1976 lecture "Beauty and the Quest for Beauty in Science," Chandrasekhar stressed how the quest for beauty—and the proper appreciation of that beauty when it was apprehended—was of fundamental importance to science, the arts and literature, supremely poetry. He opened this lecture with a quotation from Henri Poincaré:

The scientist does not study nature because it is useful to do so. He studies it because he takes pleasure in it; and he takes pleasure in it because it is beautiful. If nature were not beautiful, it would not be worth knowing, and life would not be worth living...I mean the intimate beauty which comes from the harmonious order of its parts, and which a pure intelligence can grasp.

Chandrasekhar's lectures merit close study, not least on account of his firm grasp of the important link between creativity in science and the arts, and the transcendent role of beauty in motivating and guiding the scientific enterprise.

Reading his works brings an important question to mind. We are so familiar with the fact that we can grasp the beauty of the world that we tend to take this for granted. Poincaré's argument suggests that this discernment of the beauty of nature makes science possible, in that it motivates the scientific quest. But the human mind could easily have been shaped in such a way that it did *not* regard the world as beautiful. There might be a fundamental tension, even a dissonance, between the beauty experienced within and the beauty observed without. It is as if there is a congruence or fundamental resonance between our sense of beauty and the beauty that is actually embodied in the natural world—almost as if we have been programmed or hardwired to recognize and respond to the beauty of the world.

The Christian doctrine of creation suggests that this fundamental resonance is no accident, but is grounded in the structures of creation itself. Our sense of wonder at the beauty of nature is thus an *indirect* appreciation of the beauty of God. Rightly perceived, nature points beyond itself. For some, nature is an end in itself, demanding to be explained and then exploited. Yet the doctrine of creation introduces a new dimension to nature—as a means by which the glory and radiance of God can be reflected toward humanity, already accommodated to its ability to discern it.

This point was made by Bonaventura (1217–74), a medieval Franciscan philosopher and theologian with a keen eye for the importance of the creation as a guide to its creator:

All the creatures of this sensible world lead the soul of the wise and contemplative person to the eternal God, since they are the shadows, echoes and pictures, the vestiges, images and manifestations of that most powerful, most wise and best first principle, of that eternal origin, light and fulness, of that productive, exemplary and order-giving Art. They are set before us for the sake of our knowing God, and are divinely given signs. For every creature is by its very nature and kind of portrayal and likeness of that eternal Wisdom.

If the world is indeed created, it follows that the beauty, goodness, and wisdom of its creator are reflected, however dimly, in the world around us. All of us have known a sense of delight at the beauty of the natural world. Yet this is but a shadow of the beauty of its creator. We see what is good, and realize that something still better lies beyond it. And what lies beyond is not an abtract, impersonal, and unknowable force, but a personal God who has created us in order to love and cherish us.

It is easy to see why Bonaventura's line of thought has found such a resonance within the Christian tradition and far beyond. Many have found that the awesome sight of the star-studded heavens or the solemn stillness of a wooded landscape evokes a sense of wonder, an awareness of transcendence, which is charged with spiritual significance. The distant shimmering of stars, however, does not itself create this sense of longing; it merely exposes what is already there. It is a catalyst for our spiritual reflections, revealing and heightening our emptiness and compelling us to ask whether and how this void might be filled. Might our true origins and destiny somehow lie beyond those stars? Might there not be a homeland from which we are presently exiled and to which we secretly long to return? Might not our accumulation of discontentment and disillusionment with our present existence really be a clandestine pointer to another land, where our true destiny lies, which is able to make its presence felt now in this haunting way? This idea can be found in many of the poems of Henry Vaughan (1621–95), perhaps most notably "The Tempest":

All things here shew him heaven . . .
Trees, herbs, flowers all
Strive upward still, and point him the way home.

Yet nature can come to be seen as an end in itself, rather than as a beau-tiful pointer to a still greater beauty. Those who insist that there is no tran-scendent dimension to nature rob nature of its deepest meaning, and humanity of the hope that this signifies. For the great early Christian theo-logian Augustine (354–430), the beautiful things of the world point to God as their creator. Augustine stresses how easy it is to mistake the sign for the thing signified, the lesser for the greater beauty, the secular for the transcendent. His insights at this point are partly grounded in his conver-sion experience, in which he discovered the reality of a God mirrored in, but not identical with, the creation. As Augustine mused upon his conver-sion and its implications, he came to realize the paradox of creation—that what was created had the potential to draw people away from its creator, as much as to lead to that creator. The beauty of creation could end up at-tracting the human mind to itself rather than pointing to the greater beauty of its creator.

Thus Augustine's true spiritual breakthrough took place only when he learned to see the beauty of God *through* the beauty of the created order:

> Those lovely things kept me away from you, things which would not even exist if it were not for you. You called and cried out, and shattered my deafness. You were radiant and gleaming, and put my blindness to flight. You blew fragrant breath towards me, and I drew in my breath and now long for you. I tasted you, and I hunger and thirst for you. You touched me, and I was inflamed with a longing for your peace.

Beauty, for Augustine, possessed the ability to draw humanity to God, through an ascent of the mind through the external beauty of nature to the final source of that beauty, which is God. To fail to discern God in the

created order is to shortsightedly attach affection to the created order itself
rather than to its creator:

> I loved beautiful things of a lower order, and I was sinking down
> to the depths. I used to say to my friends: "Do we love anything
> other than that which is beautiful? Then what is a beautiful object?
> And what is beauty? What is it that charms us and attracts us to
> the things that we love? It must be the grace and loveliness which is
> inherent in him; otherwise they would in no way draw us to them."

As C. S. Lewis pointed out, if we fail to realize how the beauty of nature
points to its transcendent goal, we are doomed to frustration and sadness:

> The books or the music in which we thought the beauty was lo-
> cated will betray us if we trust to them; it was not in them, it only
> came *through* them, and what came through them was longing.
> These things—the beauty, the memory of our own past—are
> good images of what we really desire; but if they are mistaken for
> the thing itself they turn into dumb idols, breaking the hearts of
> their worshippers. For they are not the thing itself; they are only
> the scent of a flower we have not found, the echo of a tune we have
> not heard, news from a country we have not visited.

The dangers of allowing the beauty of the world to displace the prior
and superior beauty of God must be noted, and allowed their proper place
in Christian reflections on nature. Yet they in no way detract from the immense significance with which they invest the natural order. To study nature
is to catch glimpses of the divine in the ordinary. It is possible to fixate
upon that sign and miss the greater reality to which it points. George Herbert put it superbly in "Teach me, my God and King" from *The Temple*:

> *A man that looks on glass*
> *On it may stay his eye;*

Or if he pleaseth, through it passe
And then the heav'n espie

The doctrine of creation invites us to value both nature and the respectful investigation of nature as a means of appreciating the splendor of the creation and glimpsing the still greater splendor of its creator. It affirms the human sense of wonder at the glories of the natural world—and hence the longing to study them more deeply—while investing them with a transcendent significance.

The Image of God in Humanity

The Christian doctrine of creation affirms that humanity is created "in the image of God" (Genesis 1:26–27). The critically important word "image" here suggests a similarity or correspondence between humanity and God, while avoiding any suggestion of an identity between them. The passage hints at an *affinity* between God and humanity, with highly significant implications for the interpretation of the age-old human longing for transcendence. We have been created purposefully; that is, we have been made to know God, in the dual sense of knowing about God, and entering into a relationship with God, as we might know a close friend. There is a natural longing within us for transcendence, which is grounded in God and which points to God.

This insight is charged with explanatory potential. Why is it that the human mind is able to discern the patterning of the world? We have already seen that this understanding of creation allows us to posit a fundamental resonance between the rationality of God, the ordering of creation, and the ability of the human mind to comprehend at least something of the rationality of the world. Why is it that there appears to be some correspondence between the rationality of the cosmos and our own rationality? If there were not, the universe would remain a mystery to us. Why is it that we are able to represent the structuring and ordering of the world in the language of mathematics, when this is supposedly the free creation of the human mind?

The answers to all these questions converge in this great truth: we have been created with the ability to peer into the mind of God. If our reasoning has its source in God, it has the potential to lead us to its fountainhead. Even though it may be attenuated through our weakness and frailty, our created reason retains its God-given ability to point us toward its creator. The stream can point us to its source. The resonance between reason, the world, and God is no accident; it is an integral aspect of the Christian doctrine of creation.

When human nature is seen through the eyes of faith, its longings and aspirations begin to make some kind of sense. If one accepts that humanity has been created "in the image of God," then there is indeed a deep God-shaped void within us—something that the French philosopher Blaise Pascal (1623–62) termed an "abyss." For Pascal, the human quest for happiness and fulfillment reflected an unacknowledged human longing for God, which was grounded in the simple fact that we are meant to relate to God. Human longing for fulfillment and despair over apparent meaninglessness make sense in the light of our true desire being for God, even if we fail to realize that this is the case.

> What else does this longing and helplessness proclaim, but that there was once in each person a true happiness, of which all that now remains is the empty print and trace? We try to fill this in vain with everything around us, seeking in things that are not there the help we cannot find in those that are there. Yet none can change things, because this infinite abyss can only be filled with something that is infinite and unchanging—in other words, by God himself. God alone is our true good.

The sense of human longing for God is thus not a delusion, to help us cope with the unbearable pain of a godless world, but a direct result of being created by God. We are made and meant to relate to God and to feel the pain of the absence of God.

This Christian interpretation of our sense of emptiness has, of course, been challenged. It is a delusion, an invention, a coping mechanism. The human mind, some argue, is superbly capable of defending itself against thoughts that it finds troubling. There is always one more portcullis to lower, another drawbridge to raise, to prevent the intrusion of threatening ideas—such as personal extinction and the meaninglessness of the cosmic void. This idea is found in the writings of both Karl Marx and Ludwig Feuerbach. There is no meaning to the world, save what we ourselves create and impose upon the world. For Feuerbach, humanity needs to find solace in consoling thoughts, such as the existence of God and heaven. Both, in Feuerbach's view, are complete fabrications, the inventions of sad and lonely souls desperately craving for meaning in a universe conspicuously devoid of purposes and goals.

Marx took this analysis a stage further and argued that the human longing for consolation was the direct result of unjust social and economic conditions. The miserable material situation of humanity was the direct cause of its spiritual yearnings. Abolish social and economic misery—for example, through a communist revolution—and people wouldn't need to dream of heaven anymore. It would have been created on earth, in a communist state.

Especially in his later writings, Marx adopted somewhat uncritically the crude progressive scientific outlook so popular in the early nineteenth century. Religion is just outmoded superstition which dulls the ability of the working class to rise up and overthrow the ruling class. It is the "opium of the people" which enables them to endure their sufferings, when they ought to be throwing off the shackles that condemned them to such suffering in the first place. We find similar ideas expressed in Lenin:

Religion is opium for the people. Religion is a kind of spiritual intoxicant in which the slaves of capital drown their humanity and blunt their desire for a decent human existence...The classconscious worker of today...leaves heaven to the priests and bourgeois hypocrites. He fights for a better life for himself, here on earth.

Once materialism had been declared to lie at the heart of official Soviet ideology, a rigidly hostile attitude to Christianity and any other religion affirming and celebrating the spiritual dimensions of existence followed as a matter of course.

Yet it can equally be argued that the mind is finely tuned to discerning signals of transcendence, patterns within the world that point to our origins and destiny lying in God. On this view, God has created us to relate to him, and if we do not do so, we lose sight of our true goal and joy. Without God, we are unfulfilled, precisely because we have been created with a God-shaped gap within us, which cries out to be filled with the luxurious presence of our creator. God has thus fashioned us in such a way that we may begin to gain at least a glimpse of his nature and being from the world around us.

This leads us on to deal with the theme of consolation, which has already been hinted at in our discussion. In his *Confessions*—a spiritual autobiography—Augustine of Hippo set out how the theme of consolation was inextricably linked with the doctrine of creation. He set this out in the form of a prayer to God: "You have made us for yourself, and our heart is restless until it finds its rest in you." For Augustine, we were created to relate to God, and our consolation is only achieved when we finally arrive in his presence. Augustine thus speaks of the present life as a time of exile—a time when we are absent from our true homeland and are sojourners in this world. It is a beautiful world, created by God, and we are required to tend it and care for it, as Adam was given the task of tending the garden of Eden. It is not our true homeland, but a place of transition. We are called to care for this earth while we look with expectation to another world, which will be like the best of this world, only better. We journey in hope, knowing that the beauty of this world is a pointer to the glory of the next and that we must cherish and preserve this world for the benefit of those who will follow us.

The doctrine of creation thus helps us to understand the deep human longing for transcendence, so important a theme in the writings of many modern philosophers. But it also allows us to understand the human sense of wonder, evoked immediately by the beauty of creation. Even before we can understand how a rainbow comes into being, we appreciate it as a thing of beauty and become aware of it awakening a longing within us for transcendence. The beauty of a mountain vista or a meadow of flowers is apprehended instantly, even before we begin to reflect on how those flowers came to be there or why they have such beautiful colors. Yet things of beauty, from a Christian perspective, evoke a *double* sense of wonder. In addition to the immediate sense of delight that their beauty evokes, we are aware of the greater beauty and glory to which they point—the ability "to behold *the sweet glory of God*" (Jonathan Edwards).

Our appreciation of a rainbow is enhanced through an understanding of the Newtonian laws of optics. This does not detract or distract from the immediate spine-tingling sense of delight at a rainbow, or from the potential of a rainbow to point beyond itself to a realm for which we can only long in our present situation, but which we believe we shall one day enter. Today it is known through our imaginations, as the bright shadow of a distant reality; tomorrow it will be known in full. Only the cruder forms of materialism that are mistakenly held to be the inevitable consequence of the natural sciences will lead to an impoverishment of the human sense of wonder at the beauty of nature. The human sense of wonder at nature is thus to be seen as a clandestine longing for God, which is both triggered and heightened by the beauty of nature.

So how do these insights relate to the ecological challenge that we now face? In what follows, we shall explore the important ecological insights that emerge from the Christian doctrine of creation.

CHAPTER TWO

Respect for Nature: Christianity and Ecological Concern

Christians see the world as God's creation, which we are called upon to "tend." This insight compels us to treat the natural world with respect, care, and concern. The breath-catching sense of wonder that we experience on encountering nature at its best is itself the symbol or sign of the deep spiritual significance of creation, which, when rightly interpreted, invites us to appreciate, honor, and respect it. This is not an idea invented to meet the needs of the moment, or a highly selective reading of a religious tradition designed to extract only those notions that happen to meet with contemporary cultural approbation. It is simply an application of a fundamental doctrine of the Christian faith to the issues we now face.

Throughout this work, a dual audience is envisaged. On the one hand, I anticipate that this work will be read by Christians of a wide variety of denominational and theological persuasions. My task here is to explore the richness of the Christian tradition and begin to make the connections between that immensely fertile and engaging tradition and the new ecological agenda. In doing so, we shall engage with some major themes from the natural sciences, the history of Western culture, literature, and philosophy.

This chapter will explore a wide range of Christian approaches to the great theme of the "tending of creation," aiming both to reassure Christians of the fecundity and resilience of their faith in this respect and also to explore how the various approaches already present within the tradition can bear on the present crisis. My task has been similar to that of the householder who brings out of his treasures "things new and old" (Matthew 13:52). Yet perhaps this exploration will do more than persuade Christians that these resources are present; it might also encourage them to be challenged and encouraged by the vision they offer, and resolve to apply them to environmental issues.

That there is a need to recover this commitment to creation is obvious. In an address to the New Delhi Assembly of the World Council of Churches in 1961, The Lutheran scholar and theologian Joseph Sittler spoke at some length of the theological importance of "the care of the earth, the realm of nature as a theater of grace, the ordering of the thick, material procedures that make available to or deprive man of bread and peace." He rejected the modernist view that nature has nothing to do with the divine, asserting that the Old Testament—to mention only one case of importance—takes a fundamentally different attitude, seeing nature as a mirror of the glory of God. Sittler's speech was warmly received but had no discernible impact. Maybe the world Christian community was not ready for such ideas back in the 1960s, when everyone was concerned about social justice, and the environment seemed an irrelevance in the face of world poverty. Things are different now.

A second audience I envision for this book has no commitment to the Christian faith but is nevertheless interested in exploring what it has to say on this matter. This audience will probably have heard the "Christianity is the enemy" mantra of some deep ecologists. Some will already be suspicious of such a simplistic dismissal of a great religious tradition, being aware of its shallow historical and intellectual foundations. Others will be interested in exploring this issue as important in itself. I do not expect these readers to share my Christian faith, but I hope that they will find the process of exploration and application interesting and informative. I hope that they will be able to view our natural environment from the vantage point that the Christian tradition offers* and appreciate its merits.

We will begin by looking at the major themes that emerge from a Christian understanding of creation. We have already explored some of its aspects; we now need to bring these strands together, to weave a tapestry emblazoned with ecological themes.

The Grand Tradition: Tending the Creation

"The fundamental relation between humanity and nature is one of caring for creation." These words, taken from the statement *Renewing the Earth*, issued by the Roman Catholic bishops of the United States in 1991, admirably summarize one of the great themes in the Christian understanding of environmental issues, as set out in the Bible and Christian reflection on this central text. The biblical notion of creation is enormously rich and complex, and offers a number of insights of determinative importance in relation to the issue of the care of creation. The following points emerge from any responsible attempt to take the biblical insights concerning creation seriously:

1. The natural order, including humanity, is the result of God's act of creation, and is affirmed to be God's possession.
2. Humanity is distinguished from the remainder of creation by being created in the "image of God." This distinction involves the delegation of responsibility rather than the conferral of privilege, and cannot be seen as offering legitimation to environmental exploitation or degradation.
3. Humanity is charged with the tending of creation (as Adam was entrusted with the care of Eden—Genesis 2:15), in the full knowledge that this creation is the cherished possession of God.

* For a very accessible and thorough introduction to Christian thought, see Alister E. McGrath, *Christian Theology: An Introduction*, 3rd ed. (Oxford: Blackwell, 2001).

4. There is thus no theological ground for asserting that humanity
    has the "right" to do what it pleases with the natural order. The
    creation is God's, and has been entrusted to humanity, which is to
    act as its steward, not its exploiter.

It is important to notice how the Christian understanding of creation
can function as the basis of a rigorously grounded approach to ecology.
This has been set out in a particularly attractive manner in a recent study by
Calvin B. DeWitt, a University of Wisconsin professor of environmental
studies who founded the Au Sable Institute for Environmental Studies. In
an important paper (1995) on religious belief as ecological practice, De-
Witt argues that four fundamental ecological principles can readily be discerned within the Christian Bible:

1. The "earthkeeping principle": just as the creator keeps and sus-
    tains humanity, so humanity must keep and sustain the creator's
    creation.
2. The "sabbath principle": the creation must be allowed to recover
    from human use of its resources.
3. The "fruitfulness principle": the fecundity of the creation is to be
    enjoyed, not destroyed.
4. The "fulfillment and limits principle": there are limits set to hu-
    manity's role within creation, with boundaries set in place that
    must be respected.

In making such basic points, it must be noted that they have generally
failed to make any real impact within the more skeptical sections of the scientific community, who persist in portraying Christianity as lending some
kind of ideological sanction to the unprincipled and unlimited exploitation
of the environment. Thus Lynn White's influential article (1967) asserted
that Christianity was to blame for the emerging ecological crisis through
using the concept of the "image of God" as a pretext for justifying human
exploitation of the world's resources. Genesis, he argued, legitimated the

notion of human domination over the creation, hence leading to its exploitation. Despite (or perhaps on account of?) its historical and theological superficiality, the paper had a profound impact on the shaping of popular scientific attitudes toward Christianity in particular and religion in general.

With the passage of time, a more sanguine estimation of White's argument has gained the ascendancy. The argument is now recognized to be seriously flawed. A closer reading of the Genesis text indicates that it endorses such themes as "humanity as the steward of creation" and "humanity as the partner of God" rather than "humanity as the lord of creation." Furthermore, a careful study of the reception of this text within the Judeo-Christian tradition makes it clear that White's interpretation cannot be sustained. The classic study of this is found in Jeremy Cohen's *"Be Fertile and Increase, Fill the Earth and Master It": The Ancient and Medieval Career of a Biblical Text* (1989), which took issue with White's suggestion that this text (Genesis 1:28) is a blatant license for environmental exploitation.

Cohen's careful study demolishes the idea that this text inculcates the "ethos which has nurtured our ecological crisis." For Cohen, the judgment of Richard Hiers—who argued that White confronted the biblical text in a "critically illiterate manner"—must be sustained. Cohen draws attention to the work of leading Old Testament scholars such as James Barr and environmental historians such as J. Donald Hughes in asserting that the "Bible blatantly lacks a spirit of ecological functionalism and technological inventiveness which Greco-Roman civilisation bequeathed to the Western tradition." In fact, biblical Israel displayed considerably more environmental sensitivity than its neighbors in the ancient Near East. After an exhaustive study of how this text has been interpreted by Jewish and Christian writers, Cohen concludes that it does not give any legitimation to environmental exploitation.

Many recent Christian writers have stressed that, far from being the enemy of ecology, the Christian doctrine of creation affirms the importance of human responsibility toward the environment. In a widely read study, the noted Canadian theologian Douglas John Hall stressed that the biblical concept of "domination" was to be understood specifically in terms of "stewardship." To put it simply: creation is not the possession of humanity; it is something that is to be seen as entrusted to humanity, which is responsible for its safekeeping and tending.

A further contribution has been made by the influential German theologian Jürgen Moltmann, noted for his concern to ensure the theologically rigorous application of Christian theology to social, political, and environmental issues. For example, in his 1985 work *God in Creation: A New Theology of Creation and the Spirit of God*, Moltmann argues that the exploitation of the world reflects the rise of technology and seems to have little to do with specifically Christian teachings. Furthermore, he stresses the manner in which God can be said to indwell the creation through the Holy Spirit, so that the pillage of creation becomes an actual assault on God. On the basis of this analysis, Moltmann is able to offer a rigorously Trinitarian defense of a distinctively Christian ecological ethic. A related approach may be found in the evangelical theologian Clark Pinnock, who comments:

> We are trustees of the earth and fellow creatures with its inhabitants. Spirit calls us to ecological consciousness. We are dependent on nature and belong to the natural order. It is not simply an object for domination and exploitation, but the Spirit's project, to be redeemed along with us. Nature is our home, blessed by God who took flesh, and it is destined for renewal.

This approach has not been invented to respond to the late-twentieth-century criticism that Christianity lacks ecological compassion and concern. The Christian tradition is replete with a deep respect for nature as God's creation. To illustrate this point, we may enter into the world of Celtic Christianity, which blossomed in Ireland, Scotland, and Wales during the seventh century, thirteen hundred years before it became fashionable to respect the environment. As we shall see, Celtic Christianity respected nature as a matter of fundamental religious principle, not in order to safeguard the future of the human race. Our concern for the latter may, however, cause us to listen to their views about the former with close attention.

The Ecological Vision of Celtic Christianity

Since about 1960, there has been growing interest in the distinctive ideas of what is loosely—and perhaps slightly misleadingly—termed "Celtic Christianity." This phrase is used to refer to the form of Christianity that flourished in Ireland during the seventh and eighth centuries, associated with figures such as Patrick, the patron saint of Ireland, and which had considerable influence far beyond. The modern interest in this movement lies partly in the Celtic celebration of nature as God's gift to humanity, and the sense of delight taken in pondering its glories and wonders. I have a particular interest in this form of Christianity, as I was baptized in the cathedral that stands on the site of St. Patrick's grave in Downpatrick, Northern Ireland.

In his influential yet fundamentally flawed article "The Historical Roots of Our Ecological Crisis," Lynn White overlooks this highly significant and influential strand of Christianity. He mentions it in passing at one point only, and in doing so, shows himself to have no appreciation of its significance for his theme. "Legends of saints, especially the Irish saints, had long told of their dealings with animals but always, I believe, to show their human dominance over creatures." This represents a serious misreading of the love and passionate concern for nature that is so typical of Celtic Christianity. White's concern to insist that Christianity encourages domination forces him to read the substantial Celtic engagement with nature in this skewed and totally misleading manner. If your only tool is a hammer, everything looks like a nail. White seems unable to see that what he believes to underscore his conviction that Christianity encourages domination is actually a counterargument that threatens his entire thesis. For Celtic Christianity, nature was God's gift to humanity, to be honored, revered, and celebrated. In many ways, it offers both a model for the modern ecological movement and a challenge to the historical inaccuracies that litter White's seminal article.

In view of its importance to our theme of the reenchantment of nature, we may consider this movement's characteristic approach to the natural world, and its contemporary implications.

Although the origins of Celtic Christianity seem to lie in Wales, it is

Ireland that established itself as a missionary center of distinction in the fifth and sixth centuries. Other centers of missionary activity in the Celtic sphere of influence are known from this period, most notably Candida Casa (modern-day Whithorn, in the Galloway region of Scotland), which was established by Bishop Ninian in the fifth century. The significance of this missionary station was that it lay outside the borders of Roman Britain and was thus able to operate without the restrictions then associated with Roman forms of Christianity.

The person who is traditionally held to be responsible for the evangelization of Ireland was a Romanized Briton by the name of Magonus Sucatus Patricius, more usually known by his Celtic name "Patrick" (c. 390–c. 460). Born into a wealthy family, Patrick was taken captive by a raiding party at the age of sixteen and sold into slavery in Ireland, probably in the region of Connaught. Here, he appears to have discovered the basics of the Christian faith, before escaping and making his way back to his family. He had been in captivity for six years. It is not clear precisely what happened between Patrick's escape from captivity and his subsequent return to Ireland as a missionary. A tradition, dating back to the seventh or eighth century, refers to Patrick spending time in Gaul before his return to Ireland. It is possible that some of Patrick's views on church organization and structures reflect firsthand acquaintance with the monasticism of certain regions of southern France. There is excellent historical evidence for trading links between Ireland and the Loire Valley around this time.

Patrick returned to Ireland and established Christianity in the region. Some form of Christianity already existed: not only does Patrick's conversion account presuppose that others in the region knew about the gospel; contemporary records dating from as early as 429 speak of one Palladius as the bishop of Ireland, indicating that at least some form of rudimentary ecclesiastical structures existed in the region. Irish representatives are also known to have been present at the Synod of Arles (314). Patrick's achievement is perhaps best understood in terms of the consolidation and advancement of Christianity rather than its establishment in the first place.

The monastic idea took hold very quickly in Ireland. Historical sources indicate that Ireland was largely a nomadic and tribal society at this time, without any permanent settlements of any importance. The monastic quest for solitude and isolation was ideally suited to the Irish way of life. Whereas in western Europe as a whole, monasticism was marginalized within the life of the church, in Ireland it rapidly became its dominant form. It is no exaggeration to say that the Irish church was monastic, with the abbot rather than the bishop being seen as preeminent.

Celtic church leaders were openly critical of worldly wealth and status, including the use of horses as a mode of transport, and any form of luxury. Instead, they advocated a life of simplicity, in close harmony with the God-given state of nature. Theologically, Celtic Christianity stressed the importance of the world of nature as a means of knowing God. This is especially clear from the ancient Irish hymn traditionally ascribed to Patrick, known as the "Deer's Cry." The hymn shows a fascination with the natural world as a means of knowing God and appreciating God's glory:

> I bind to myself today
> The power of Heaven,
> The light of the sun,
> The brightness of the moon,
> The splendour of fire,
> The flashing of lightning,
> The swiftness of wind,
> The depth of sea,
> The stability of earth,
> The compactness of rocks.

The Celtic vision of nature goes further than a mere respect for the natural order. There is a strong sense of a reciprocal relationship between God and the creation. By living so close to nature, the Celtic Christians could not but be overwhelmed by the sense of the presence of God in nature, which seemed to proclaim and extol its creator at every turn. As a ninth-century Celtic poem put it:

Almighty Creator, who has made all things
The world cannot express all your glories,
Even if the grass and the trees were to sing.

Animals, birds, plants, and running water were all seen as tangible intimations of divine glory. Many Celtic monasteries were situated in remote regions of Ireland, close to nature. The monks had ample time to study nature and came to love its rich tapestry. The illustrations in Celtic manuscripts show a concern for the fine details of the natural world. This fascination with the natural order of things was grounded in a belief that it hinted at the greater glory of God and was a way of drawing closer to that glory. As a work attributed to Ninian of Whithorn argued, the supreme aim of the study of nature is "to perceive the eternal word of God reflected in every plant and insect, every bird and animal, and every man and woman." Others echoed this theme, even if their names have been lost to history.

There is no life in the sea,
No creature in the river,
Nothing in the heavens,
That does not proclaim God's goodness

There is no bird on the wing,
No star in the sky,
Nothing beneath the sun,
That does not proclaim God's goodness.

The environmental implications of this will be obvious. Celtic Christianity of the seventh, eighth, and ninth centuries could know nothing of the environmental crisis of today. Indeed, such was the respect in which the movement held nature that such a crisis would have been inconceivable to them. How could anyone fail to love and respect nature? The Celtic tradition encourages us to value nature for what it is in itself and for what it reminds us of and what it proclaims. As God's good creation, it is to be honored and cherished. As the Irish writer Columbanus (543–615) put it, "if you want to know the creator, understand the created things."

A passionate love and concern for nature is deeply embedded within this form of Christianity, calling out to be reappropriated by its modern successors. The task facing the Christian church today is thus to regain a spiritual insight that seems to have been neglected within Western Christianity yet was taken for granted by our forebears. As will become clear from what follows, many have found this process of retrieval to be both spiritually satisfying and ecologically fulfilling.

Christian Responses to the Environmental Crisis

We have already seen how a deep love and respect for nature is characteristic of the Christianity that developed many centuries ago. There has been no shortage of comparable affirmations of concern and respect from Christian churches in recent years. Those ecologists who continue to insist woodenly that "Christianity is the enemy" could do a lot worse than read the myriad of statements on environmental issues by Christian denominations, particularly in the United States, in the last three decades. For example, in June 1989, the American Baptist Churches of the United States urged their members to:

1. Affirm the goodness and beauty of God's creation.
2. Acknowledge our responsibility for stewardship of the creator's good earth.
3. Learn of the environmental dangers facing the planet.
4. Recognize that our practices and styles of life have had an effect on the environment.
5. Pursue a lifestyle that is wise and responsible in light of our understanding of the problems.
6. Exert our influence in shaping public policy and insisting that industries, businesses, farmers, and consumers relate to the environ-

ment in ways that are sensible, healthy, and protective of its integrity.

7. Demonstrate concern with "the hope that is within us," as despair and apathy surround us in the world (Romans 12:21).

8. Become involved in organizations and actions to protect and restore the environment and the people in our communities.

A number of major themes can be discerned within these important statements, each reflecting a fundamental Christian insight into the place and responsibilities of humanity within God's creation.

1. The sinfulness of humanity.

The Christian tradition stresses that humanity is sinful, having departed from or fallen short of what God intended for it. The greed, self-centeredness, and exploitative traits we see in human individuals and societies have important implications for the environment, as the Episcopal Church of the United States has emphasized. This church established its Peace and Integrity of Creation Committee in February 1995 to continue the work of bodies created by previous general conventions or committees of the Executive Council. These include the Economic Justice Implementation Committee, the Environmental Stewardship Team, Jubilee Ministries, and the Racism Commission. The committee stressed the importance of recognizing the devastating impact of human sin upon the global environment and called upon the Episcopal Church to be a witness of hope and bearer of judgment on this issue:

Our church is becoming the light in a great darkness. The earth lies polluted under its inhabitants, for they have transgressed laws, violated the statutes, broken the everlasting covenant. "Therefore a curse devours the earth and its inhabitants suffer for their guilt" (Isaiah 24:5–6). We are living in a time when the shepherd of today, the church, must guide the blindly following sheep away from running themselves over the cliff.

"If my people which are called by my name will humble themselves and pray and seek my face and turn from their wicked ways, then I will hear from Heaven. I will forgive their sins and heal their land." (2 Chronicles 7:14)

Greed must be healed. The economics which drive creation's destruction, the dumping of toxic waste and garbage on minority communities, the devastation of forests and wetlands, the total disregard for every living thing, and the inability to find peace in our lives comes from greed. Violence is as subtle as pollution and as horrific as murder. Pollution of our planet affects the health of every living thing. We must learn that violence is the barometer by which we see the manifestation of our spirituality.

2. The land is God's, not ours.

The doctrine of creation affirms that the created order is to be respected as a gift of God. Humanity does not own the land or have rights to exploit it as it pleases. This point is made clearly in a statement issued by the United Methodist Church in 1988. The statement was prompted by the specific crisis faced by rural American communities at that time, set alongside a more general growing awareness of the serious implications of environmental degradation.

> God is the owner of the land (Leviticus 25); thus it is a gift in covenant which involves the stewardship of keeping and tending the land for present and future generations; as God's creation, land has the need to be regenerated that it may sustain life and be a place of joy. It is a common gift to all of life requiring just patterns of land use.
>
> Social, economic, and ecological justice with regard to the use of land was central to the Law. The land itself was to receive a rest every seven years (Leviticus 25:4). Voluntary charity or occasional care of the land was not enough. Israel's failure to follow the laws related to the land was considered a cause of the exile to Babylon (2 Chronicles 36:21).

The care of the land, the rights of the poor and those in need were at the center of the Law. Adequate food was regarded as an inherent right of all, such that the poor could eat grapes in a neighbor's vineyard or pluck grain when passing by a field (Deuteronomy 23:24–25). Owners were urged not to be too efficient in their harvest (Leviticus 19:9–10), so that gleaning by those in need was possible.

3. Violating nature is to be seen as a sin.

The seriousness of violating the environment was stressed in an address given by the ecumenical patriarch Bartholomew at the environmental symposium held at the St. Barbara Greek Orthodox Church in Santa Barbara, California, on November 8, 1997. This conference took place against the backdrop of an oil spill which threatened to pollute local beaches. The patriarch affirmed the traditional Greek Orthodox view that it was necessary to see human "life and the world as a sacrament of thanksgiving" and that the traditional Orthodox virtue of asceticism—that is, self-discipline—was essential if humanity was to correct its flawed attitude to nature. "Asceticism requires from us a voluntary restraint, in order for us to live in harmony with our environment." Perhaps the most significant aspect of the address was its firm stipulation that deliberate degradation of the environment was sinful—that is, something to which God was opposed and which required repentance and amendment of life.

> If human beings treated one another's personal property the way they treat their environment, we would view that behavior as antisocial. We would impose the judicial measures necessary to restore wrongly appropriated personal possessions. It is therefore appropriate for us to seek ethical, legal recourse where possible, in matters of ecological crimes.
>
> It follows that, to commit a crime against the natural world is a sin. For humans to cause species to become extinct and to destroy the biological diversity of God's creation ... for humans to degrade the integrity of Earth by causing changes in its climate, by

stripping the Earth of its natural forests, or destroying its wetlands . . . for humans to injure other humans with disease . . . for humans to contaminate the Earth's waters, its land, its air, and its life, with poisonous substances . . . these are sins.

In prayer, we ask for the forgiveness of sins committed both willingly and unwillingly. And it is certainly God's forgiveness, which we must ask, for causing harm to His Own Creation.

In every case, the denominational response to these important environmental issues is not something dreamed up on the spur of the moment, representing an ad hoc response to issues, but is to be seen as the application of the fundamental themes of the Christian tradition to the problem at issue. Having outlined some representative responses to the ecological agenda, we may consider three in more detail. These have been chosen to reflect significantly different elements in modern Western Christianity, to indicate how the movement as a whole is firmly committed to caring for the earth.

Roman Catholicism

Roman Catholicism is by far the world's largest Christian denomination and represents the most numerous religious group in the United States, with four times the membership of its nearest rival, the Southern Baptist churches. Recent statistics suggest a modest growth in its membership over the past few years. It is therefore of considerable importance to examine how this highly influential Christian group has engaged with environmental issues.

The potential contribution of Catholicism to the environmental debate can be seen from the landmark document "The Columbia River Watershed: Caring for Creation and the Common Good" (February 2001). This took the form of an eighteen-page pastoral letter from twelve Catholic bishops, representing 1.5 million Catholics in the Northwest United States and Canada, in which the bishops declared that the Columbia River—the longest river in North or South America flowing into the Pacific Ocean—is threatened with environmental degradation.

The Columbia River has one of the biggest drainage basins on the continent and serves an ecosystem of 259,000 square miles (650,000 square kilometers). The river has its source in Canada's Rocky Mountains. The first third of its 1,200-mile course meanders through the Canadian province of British Columbia before entering the states of Washington and Oregon. A major tributary, the Snake River, flows through Washington and Idaho, with its headwaters in Wyoming. There are fourteen dams on the two rivers, eleven of them in the United States.

In their pastoral letter, the twelve bishops argued that the river had been irresponsibly dammed, polluted, and overfished, and called on all people of goodwill to "work together to develop and implement an integrated spiritual, social and ecological vision for our watershed home, a vision that promotes justice for people and stewardship of creation."

The document is important because of the way in which it fuses genuine environmental concern with a solid grounding in the Christian doctrine of creation, making connections between what is believed and trusted, and what needs to be done in the world. The letter affirms that:

> Stewardship is the traditional Christian expression of the role of people in relation to creation. Stewards, as caretakers for the things of God, are called to use wisely and distribute justly the goods of God's earth to meet the needs of God's children. They are to care for the earth as their home, and as a beautiful revelation of the creativity, goodness and love of God. Creation is a "book of nature" in whose living pages people can see signs of the Spirit of God present in the universe, yet separate from it.
>
> The individual members of the human family are called to respect both creation and Creator and are responsible for that part of the earth entrusted to their stewardship, whether by property rights or managerial responsibility. They are to take care of the earth out of respect for the Creator who loves all creatures, and out of a charity that calls us to love our neighbor.

The watershed document can be seen as embodying the general themes set out a few years earlier by the United States Catholic Conference and applying them to a specific situation of need and concern. That conference insisted that humans bear "a unique responsibility under God: to safeguard the created world and by their creative labor even to enhance it." Humans have a responsibility toward the creator for their use of the creation and are not at liberty to do with it as they please. Furthermore, the conference points out that "as faithful stewards, fullness of life comes from living responsibly within God's creation." It is important to appreciate that the Christian tradition does not hold that raw nature, undisturbed by human intervention, constitutes God's intended pattern for things. Humanity is privileged in being the climax of that creation. Yet with that privilege to live within the natural order comes a responsibility to care for that order.

Evangelicalism

Evangelicalism—the form of Christianity I myself adopt—can be traced back to medieval Europe, although its modern forms date from the eighteenth-century evangelical revivals in England. John Wesley (1703–91) and his brother Charles (1707–88) pioneered a renewal movement within the Church of England, which eventually led to their being thrown out of that church for excessive "enthusiasm." Both Wesleys believed that many of their colleagues within the English national church were but "half-Christians," who had no serious emotional or personal commitment to their faith. A renewal of both heart and mind was required.

Evangelicalism became increasingly important in British religious life in the late eighteenth and early twentieth centuries. Emigration from Britain to the United States led to the movement developing there, virtually achieving the status of a folk religion in many parts of the country, particularly the southern states. However, its recent history owes much to the growing reaction against fundamentalism in the late 1940s. Many conservative Protestants who were sympathetic to the aims of the fundamentalist movement became alienated by its belligerence, anti-intellectualism, and cultural separatism. Surely there was a way in which the basic beliefs of the movement could be articulated in a more sensitive, intelligent, and culturally interactive manner? Billy Graham rapidly became a figurehead of a new movement, initially called neo-evangelicalism and then simply evangelicalism. Institutions such as Wheaton College and Fuller Theological Seminary and journals such as *Christianity Today* became flagships of the new movement, which rapidly gained momentum and influence.

Evangelicalism began to emerge as a movement of major public importance in the United States during the 1950s. Full public recognition in America of the new importance and visibility of evangelicalism is generally thought to date from the early 1970s. The crisis of confidence within American liberal Christianity in the 1960s was widely interpreted to signal the need for the emergence of a new and more publicly credible form of Christian belief. In 1976, *Newsweek* magazine informed its readers that they were living in the "Year of the Evangelical," with a born-again Christian (Jimmy Carter) as their president. The result was an unprecedented media interest in evangelicalism.

The British scholar David Bebbington has provided a definition of evangelicalism that has gained wide acceptance. According to Bebbington, evangelicalism is basically a form of orthodox Christianity that possesses four distinctive hallmarks:

1. *Conversionism*—the belief that lives need to be changed through the personal appropriation of faith. A biblical text that is often cited in this context by evangelical preachers, such as Billy Graham, is "you must be born again" (John 3:7).

2. *Activism*—the actualization of Christian faith in life, particularly in evangelism (the preaching of the gospel to others) and other forms of Christian activity. One of the reasons that so many evangelical churches are so successful is that their memberships tend to be very active in outreach and discipleship programs.

3. *Biblicism*—a focus on the Bible as the most fundamental resource for Christian life and thought. Bible study is often at the heart of evangelical spiritual life, both individual and corporate. A surefire indicator of this trait is the enormous number of devotional and academic works of biblical scholarship produced by evangelical publishing houses in an attempt to meet this huge demand from their constituency.

4. *Crucicentrism*—a focus on the cross of Christ, and the benefits this brings to humanity. Many evangelical hymns take the form of meditations on the cross, such as George Bennard's "The Old Rugged Cross" and Isaac Watts's "When I Survey the Wondrous Cross."

The surge in growth in evangelicalism in the West during the past half century can be illustrated in many ways. An excellent example is provided by considering theological education in Canada. In 1950, the dominant theological schools or seminaries were owned by the mainline denominations. Over the next decades, these tended to merge, producing collaborative institutions such as the Vancouver School of Theology, Atlantic School of Theology, and so on. Although interdenominational, these schools were generally able to ensure that each denomination's interests were firmly and fairly represented. In 1992, however, the two largest theological schools were Regent College and Vancouver and Ontario Theological Seminary— both evangelical institutions founded in the 1970s. Evangelicalism had eclipsed its rivals. It is thought that about 30 percent of the population of the United States is now affiliated with evangelical denominations or congregations.

Evangelicalism is perhaps the form of Christianity that has had least interest in the environment in the recent past. Suspicious of anything that smacks of the "social gospel," the movement has generally been wary of dealing with social issues, including the environment. Yet all that has now changed. The 1994 "Evangelical Declaration on the Care of Creation" brought together many of the leading representatives of the movement in a common affirmation of the legitimacy of ecological concerns for evangelicals. The document affirms the fundamental theological linkage between the creation of humanity and the world by God, and the responsibility of humanity toward God for the creation:

Because we worship and honor the Creator, we seek to cherish and care for the creation.

Because we have sinned, we have failed in our stewardship of creation.

Therefore we repent of the way we have polluted, distorted, or destroyed so much of the Creator's work.

Evangelicalism here reiterates many of the themes we discern in this book—most notably, the importance of the theme of creation as the theological foundation of a Christian approach to the environment. The declaration makes four central statements:

As followers of Jesus Christ, we believe that the Bible calls us to respond in four ways:

First, God calls us to confess and repent of attitudes which devalue creation, and which twist or ignore biblical revelation to support our misuse of it. Forgetting that "the earth is the Lord's," we have often simply used creation and forgotten our responsibility to care for it.

Second, our actions and attitudes toward the earth need to proceed from the center of our faith, and be rooted in the fullness of God's revelation in Christ and the Scriptures. We resist both ideologies which would presume the Gospel has nothing to do with the care of non-human creation and also ideologies which would reduce the Gospel to nothing more than the care of that creation.

Third, we seek carefully to learn all that the Bible tells us about the Creator, creation, and the human task. In our life and words we declare that full good news for all creation which is still waiting "with eager longing for the revealing of the children of God" (Romans 8:19).

Fourth, we seek to understand what creation reveals about God's divinity, sustaining presence, and everlasting power, and what creation teaches us of its God-given order and the principles by which it works.

This important statement indicates how evangelicalism has firmly committed itself to recognizing the importance of caring for the creation. A movement that prides itself on being grounded and guided by the Bible discovered and applied its rich ecological resources. As evangelicalism continues to grow, it is to be hoped that environmental issues will be given an even higher profile in years to come.

Liberal Protestantism: Process Thought

One of the most important movements in liberal Protestant writings since World War II is known as process thought or process theology. This way of thinking about the relation between God and the world has been taken up by a number of prominent Protestant writers, especially those interested in the relation between Christian theology and the natural sciences, such as Ian Barbour. The origins of process thought are generally agreed to lie in the writings of the Anglo-American philosopher Alfred North Whitehead (1861–1947), especially his important work *Process and Reality* (1929). Whitehead's process philosophy marks a radical break with the scientific materialism of the late nineteenth century and a willingness to reconsider the interconnectedness and subjectivity of nature.

Reacting against the rather static view of the world that he associated with traditional metaphysics (expressed in ideas such as "substance" and "essence"), Whitehead conceived reality as a process. The world, as an organic whole, is something dynamic, not static. It is something that *happens*. Reality is made up of building blocks of what Whitehead termed "actual entities" or "actual occasions" and is therefore to be thought of in terms of becoming, change, and event.

All these entities or occasions (to use Whitehead's original terms) possess a degree of freedom to develop and be influenced by their surroundings. It is perhaps at this point that the influence of biological evolutionary theories can be discerned: like the later writer Pierre Teilhard de Chardin, Whitehead is concerned to allow for development within creation, subject to some overall direction and guidance. This process of development is thus set against a permanent background of order, which is seen as an organizing principle essential to growth. Whitehead argues that God may be identified with this background of order within the process. Whitehead treats God as an "entity," but distinguishes God from other entities on the grounds of imperishability. Other entities exist for a finite period; God exists permanently. Each entity thus receives influence from two main sources: previous entities and God.

Causation is thus not a matter of an entity being coerced to act in a given manner: it is a matter of *influence* and *persuasion*. Entities influence each other in a "bipolar" manner—mentally and physically. Precisely the same is true of God, as for other entities. God can only act in a persuasive manner, within the limits of the process itself. God "keeps the rules" of the process. Just as God influences other entities, so God is also influenced by them. God, to use Whitehead's famous phrase, is "a fellow-sufferer who understands." God is thus affected and influenced by the world.

So how do these ideas have any relevance to the ecological themes of this book? The answer lies in Whitehead's basic insistence that it is reality as a whole, rather than any particular section or element of that reality, that is to be thought of as living and active. Whitehead argued that all life is driven by a threefold urge: "to live, to live well, and to live better." According to Whitehead, all the basic constituents of reality at every level must be thought of as possessing a lifelike quality, right down to subatomic particles. Reality consists of organisms, pulses of life aiming at satisfaction. Therefore, the world is to be thought of as being made up of nothing but experiencing subjects who are capable of enjoyment.

Most process thinkers would argue that the value of organisms increases as the range and depth of their capacity for enjoyment increases. Value is thus measured in terms of the richness, complexity, and intensity of feeling that individual organisms are able to experience. In that human beings have a much wider and deeper capacity for experience and enjoyment than plants, it is argued that humans have a greater intrinsic value than plants. Nevertheless, the suggestion that individual elements of creation may have greater value than others must be set alongside the fundamental idea that *all* living beings have *some* intrinsic value.

The relevance of this approach to ecology will be clear. Process thought lays the foundations for the idea of the world as an aggregate of interacting sentient entities, each of which has individual value. The "process" as a whole is to be respected and valued, rather than those specific aspects of it for which we may feel particular sympathy. Above all, process thought marks a decisive rejection of the great reductionist model of modernity—that nature is a "mechanism" or "machine"—that has played such a decisive role in encouraging the exploitation of nature. We shall consider this idea in more detail later in this work.

The Restoration of Eden: Hope for Creation

Our attention thus far has tended to focus on the doctrine of creation and its implications for Christian ecological reflection. Yet it would be unfair to suggest that this is the sole—if highly significant—contribution that Christian theology has to make to humanity's struggle to respect and protect the environment.

One aspect of Christian theology that might be considered here is its doctrine of God. Christianity is distinguished by a rich and complex notion of God, which defies reduction to the simplistic categories of the philosophy of religion. The doctrine of the Trinity is perhaps the most puzzling aspect of Christian theology, seeming at times to be a piece of faulty logic that rests on some bad celestial mathematics. Yet the real purpose of the doctrine of the Trinity is to ensure that the Christian experience of God is not impoverished. This point is made particularly clearly in *God in Creation: A New Theology of Creation and the Spirit of God* by Jürgen Moltmann:

> The trinitarian concept of creation binds together God's transcendence and his immanence. The one-sided stress on God's transcendence in relation to the world led to deism, as with Newton. The one-sided stress on God's immanence in the world led to pantheism, as with Spinoza. The trinitarian concept of creation integrates the elements of truth in monotheism and pantheism.

Moltmann points out that the Christian doctrine of the Trinity affirms what is good and true in alternative visions of the nature of God, without being limited by their weaknesses and failings. Deism holds that God, having created the world, is no longer involved in its affairs, and leaves it, untended, to its own fate and devices. (Incidentally, this is the impoverished view of God that seems to underlie Lynn White's 1967 article.) Pantheism reduces God to a life force within the world, forbidding us to speak of it in personal terms. The Trinitarian conception of God affirms that God is to be thought of as both creator of the world and a creative presence within it. And if God inhabits the natural order, the place of the divine habitation must be treated with respect. In the Christian tradition, nature is not divine, nor is it possessed of any divine qualities. Nevertheless, it is and remains both God's possession and the place of indwelling of the one "who fills all in all" (Ephesians 1:23).

The Christian idea of the natural order as God's place of action and dwelling is intensified by the doctrine of the incarnation, perhaps one of the most remarkable Christian ideas. In essence, the doctrine holds that God did not choose to remain in heaven, but entered into human history in the form of a human being. Rather than demanding that we ascend to God in order to be saved, God chose to enter into our world, to meet us there and bring us home. In insisting that Jesus Christ is both divine and human, Christian theologians affirm that God entered into the natural world and redeemed it from within. If God valued this world enough to enter into it, and dignify it with the divine presence, then Christians ought to hold that place of habitation with appropriate respect.

Our main concern, however, in this section is with the theme of the *restoration of creation*—the Christian affirmation that there will be a final act of consummation, in which the existing order will be renewed and refashioned to create a "new heaven and a new earth" (Revelation 21:1). This contrasts with the vision of the future that Richard Dawkins offers in his 1996 article in the *Humanist,* in which he sets out how science offers "uplift" to the human soul. "The merest glance through a microscope at the brain of an ant or through a telescope at a long-ago galaxy of a billion worlds is enough to render poky and parochial the very psalms of praise." Yet Dawkins goes on to offer a vision of the future in which the universe irreversibly decays. "We know from the second law of thermodynamics that all complexity, all life, all laughter, all sorrow, is hell-bent on leveling itself out into cold nothingness in the end. They—and we—can never be more than temporary, local buckings of the great universal slide into the abyss of uniformity." The evolutionary struggle on earth will continue until the day when the "sun will engulf the earth." The tension between Dawkins's "uplift" at the sight of nature and his cosmic pessimism concerning where reflection on nature leads us is striking, and not satisfactorily resolved within his writings.

The Christian vision of the future takes the form of the renewal and transformation of creation. The day will come when "creation itself will be set free from its bondage to decay" (Romans 8:19–23) and achieve the glorious freedom for which it was created. This theme is reiterated throughout the Old Testament, which frequently looks forward to the final restoration and reintegration of nature. What has been distorted and ruined will finally be restored to its original integrity—including the animal kingdom: "The wolf and the lamb shall feed together, the lion shall eat straw like the ox; but the serpent—its food shall be dust! They shall not hurt or destroy on all my holy mountain, says the Lord" (Isaiah 65:25).

In one of his most passionate sermons, John Wesley drew out the implications of such passages for the Christian understanding of heaven. Heaven is not simply about paradise restored, but about paradise transcended. The "enormous bliss of Eden" (John Milton) will be exceeded, through a renewal and perfection of the original creation. Creation will be renewed and restored:

> The whole brute creation will then, undoubtedly, be restored, not only to the vigour, strength, and swiftness which they had at their creation, but to a far higher degree of each than they ever enjoyed. They will be restored, not only to that measure of understanding which they had in paradise, but to a degree of it as much higher than that, as the understanding of an elephant is beyond that of a

worm. And whatever affections they had in the garden of God, will be restored with vast increase; being exalted and refined in a manner which we ourselves are not able to comprehend. The liberty they then had will be completely restored, and they will be free in all their motions. They will be delivered from all irregular appetites, from all unruly passions, from every disposition that is either evil in itself, or has any tendency to evil. No rage will be found in any creature, no fierceness, no cruelty, or thirst for blood. So far from it that "the wolf shall dwell with the lamb, the leopard shall lie down with the kid, the calf and the young lion together; and a little child shall lead them. The cow and the bear shall feed together, and the lion shall eat straw like an ox. They shall not hurt nor destroy in all my holy mountain" (Isaiah 11:6–9).

The theme of the renewal of creation has important implications for our present attitude toward that creation. Francis Schaeffer, an influential evangelical writer with a particular concern for cultural issues, spoke prophetically in 1970 in making this connection:

> On the basis of the fact that there is going to be total redemption in the future, not only of man but of all creation, the Christian who believes the Bible should be the man who—with God's help and in the power of the Holy Spirit—is treating nature now in the direction of the way nature will be then. It will not be perfect, but it must be substantial, or we have missed our calling . . . we should exhibit a substantial healing here and now, between man and nature and nature and itself, as far as Christians can bring it to pass.

Schaeffer's point is simple: our attitude toward nature *now* must be informed by our knowledge of what nature is finally going to be. Redemption is an immensely rich idea for Christians, embracing a wide range of themes, including personal transformation, a renewed relationship with God, and the hope of eternal life—and also the restoration of the creation, according to God's original intentions. If the Christian vision of paradise includes the theme of the renewal of creation, we are confronted with a new motivation for ecological action—the need to preserve what will one day be a new paradise.

Perhaps Christians have been slow to realize the full ecological implications of their rich theological heritage. Yet there can be no doubt that Christianity possesses and is distinguished by a set of beliefs that affirm the importance of respecting, tending, and preserving the natural order. There are doubtless many "bad" Christians, who fail to appreciate what their tradition demands of them, or who prefer to overlook the implicit ecological dimensions of their faith. That is, however, a criticism of individual Christians, not of the fundamental vision of Christianity itself. I have no doubt that Christians need to be more attentive and sensitive to this issue, and to welcome criticism of individuals and churches when they fail to live up to their ideals in these matters. This is a vitally important role that environmentalists from outside the Christian faith can play in keeping the churches faithful to their calling. But it is quite untrue to suggest that Christianity itself, by definition and on account of its fundamental ideas and values, is antienvironmental.

So if Christianity cannot reasonably be held to be guilty for our ecological crisis, who is? Paradoxical though it may appear at first sight, there is an excellent case to be made for our problems arising from a deliberate decision to *reject* the idea of God in order to promote human freedom. Without God, humanity must no longer work under authority and under limits, but is free to do as it pleases. The critical role of Christianity in emphasizing human accountability for the environment, and placing limits on human exploitation of nature, is only now being recognized—at a time when it is needed more than ever. We shall explore this matter further in the following chapter.

CHAPTER THREE

A Manifesto to Exploit: The Enlightenment and the Master Race

 Go forth and dominate nature! For some, such as Lynn White, this is what Christianity is all about. Those who know Christianity well are not a little surprised by this, not least because the Book of Genesis—to which White appeals rather selectively for his interpretation of Christianity—lays down that humanity is to "tend" its Eden. This is a gentle image of care and compassion, suggesting that the same care that God demonstrates in loving the world is to be reflected in our attitudes toward it.

Yet, as we shall see, there is a pervasive trend in Western culture that does indeed explicitly see its mission as to "go forth and dominate nature!" There is indeed a movement that holds that human liberation and fulfillment come about through the domination of the natural world. But it is not Christianity. In what follows, we shall identify this movement and explore its implications.

The Rise of Anthropocentrism

Lynn White is completely right when he argues that human self-centeredness is the root of our ecological crisis, and completely wrong when he asserts that "Christianity is the most anthropocentric religion the world has seen." The most self-centered religion in history is the secular creed of twentieth-century Western culture, whose roots lie in the Enlightenment of the eighteenth century and whose foundation belief is that humanity is the arbiter of all ideas and values. Steven Vogel makes this clear in *Against Nature: The Concept of Nature in Critical Theory*, a careful study of the way in which the Enlightenment "disenchants" nature:

> Enlightenment is marked by the "disenchantment" of nature, its transformation from something sacred into mere matter available for human manipulation ... The project of enlightenment aims above all at the *domination of nature*. Disenchanted and objectified nature, appearing now in the guise of meaningless matter, is seen by enlightenment simply as something to be overcome and mastered for human purposes, and not to be imitated, propitiated, or religiously celebrated.

Lying behind this desire to dominate nature is the belief that humanity is of central and defining importance. Humanity is the creator and arbiter of values and is free to interpret and manipulate nature as it pleases. When this anthropocentric worldview is provided with tools that enable it to achieve its goal of dominating nature—instead of merely dreaming about it—then the environment is really in trouble. The roots of our ecological crisis lie in the rise of a self-centered view of reality that has come into possession of the hardware it needs to achieve its goals. Yet technology did not just "happen." For many scholars, the development of technology represents the outcome of a purposeful and sustained human quest, fueled by a self-centered ethic, for the means necessary to achieve the goal of the domination of nature.

Yet the dream of dominating nature is most emphatically not a Christian idea; indeed, it could be argued to be fundamentally antithetical to the central ideas of the Christian faith. So where *did* it come from? The answer lies in the revival of the classic idea that the world exists to serve the needs of humans, who are free to treat it as they please. The idea has its origins in classical Greek philosophy, was eclipsed through the rise of Christianity, and enjoyed a resurgence from the sixteenth century onward. It is not to the Christian faith that we should look, but to one of its spiritual and intellectual rivals, for the origins and development of the environmentally harmful ethos we have inherited.

In one of Plato's dialogues, the philosopher Protagoras declares his belief that "man is the measure of all things." Ideas and values were judged according to whether they suited and served human beings. We can see here the origins of an idea that is deeply interwoven into the fabric of secular Western thought—that all things are there to serve the needs of humanity. We are free to choose to behave as we like rather than have ideas or values forced upon us by others. Making the rules is to be seen as the ultimate expression of human freedom. *To limit is to enslave.*

This thread of thought can be found throughout the complex pattern of Western intellectual history. During the Middle Ages, it was relegated to the background, displaced by the dominant Christian idea that there was some intrinsic ordering to nature which could be discerned and which was to be respected. God had made nature, and there were limits to what humans were allowed to do with it. Humans saw themselves as being part of a greater cosmic ordering, which offered them both privileges and responsibilities. Human beings might well stand at the apex of nature; they did not, however, have the right to alter its course or change its contours. (The medieval hostility to mining is especially important at this point: while some saw this as an easy way of making money, others saw it as the disfigurement of the God-given face of nature.)

With the birth of the Renaissance, the ideas of classical philosophy began to enjoy a new popularity. The Renaissance had its origins in Italy and is usually traced back to the fourteenth century. It saw the culture of ancient Greece and Rome as setting norms that all civilized people should emulate. To be cultured was to soak up the ideas and values of the ancient world—its style of architecture, its way of speaking, and its ways of think-ing. The great cultural goal became the "pursuit of eloquence"—the seri-ous attempt to replicate the wisdom of the ancient world. Its impact was felt at every level of society. Gothic architecture was displaced by the ele-gant classical facades of Palladio. The writings of classic authors such as Cicero and Plato were studied with a new earnestness. The new literary cul-ture of the Renaissance modeled itself on both the style and the ideas of these writers. The theme of "man as the measure of all things" began to play an increasingly important role in Western thought and culture.

The idea finally came into its own during the period of Western cul-ture known as the Enlightenment, which is usually held to date from about 1750. As Princeton philosopher Jeffrey Stout has shown, the roots of the Enlightenment lie in a "flight from authority"—a deliberate and principled rejection of the authority of anything and anyone other than individual hu-man reason. A revolt against God, the leading thinkers of the Enlighten-ment argued, liberated humanity from the tyranny of the church and bondage to outmoded superstitions. A revolt against the past set people free from slavery to the outdated habits of thinking and behaving, which locked them into patterns of oppression. It was time to start all over again. Human reason, Enlightenment thinkers believed, was endowed with all the resources necessary for the education, ennobling, and advancement of the human race.

This emphasis upon the supreme authority of human reason proved to be an excellent way of liberating society from influences that Enlighten-ment rationalism judged to be inappropriate—such as the baleful influence of religion, and the dead hand of the past. Some Enlightenment thinkers—such as G. E. Lessing—judged religion to be harmless and were prepared to tolerate it, providing it acknowledged the authority of reason. Others—such as Voltaire and Denis Diderot—regarded it as debasing and oppressive and took the view that its forcible elimination offered the best hope for hu-manity. This latter view would triumph in the heady early years of the French Revolution of 1789.

The French Revolution was welcomed by freethinkers across Europe as marking the dawn of a new era. In 1804, the young English poet William

Wordsworth penned words that encapsulated this sense of optimism and hope among the youth of Europe. The French Revolution had shattered the tired old political framework of Europe, sweeping away its outdated tradition-bound practices and beliefs, and opening the way to a bright new future.

> *Bliss was it in that dawn to be alive*
> *But to be young was very heaven!*

Wordsworth's feelings were echoed by many young people throughout western Europe. Here, at long last, was something new, something *liberating*, which a repressed and disillusioned youth could embrace. The future seemed to belong to them. The old taboos, superstitions, and traditions would be swept aside, leaving people free to construct a brave new world in whatever way they pleased.

This ethos pervaded the Victorian period, which believed that unprecedented social and technological advances went hand in hand. The Victorian adulation of the achievements of science rested on the firm belief that the mastery of nature would solve human problems. Winwood Reade's highly influential 1872 work *The Martyrdom of Man*—much admired by writers such as Arthur Conan Doyle, H. G. Wells, and George Orwell—reflects the uncritical evolutionary optimism characteristic of this age, as it charts a vast historical panorama setting out the inexorable rise of man from one who was oppressed by nature and religion to one who had finally mastered both.

> When Man first wandered in the dark forest, he was Nature's serf
> . . . But as time passed on, he ventured, to rebel; he made stone his
> servant; he discovered fire and vegetable poison; he domesticated
> iron; he slew the wild beasts or subdued them; he made them feed
> him and give him clothes. He became a chief surrounded by his
> slaves . . . The river which once he had worshipped as a god, or
> which he had vainly attacked with sword and spear, he now con-
> quered to his will.

Central to Reade's thesis is that nature must be disenchanted, evacuated of any concept of spiritual or religious significance, before it can be harnessed to human progress and the advancement of civilization. To understand nature is to identify its weaknesses and hence to enslave it for human ends.

> We can conquer Nature only by obeying her laws, and in order to obey her laws we must first learn what they are. When we have ascertained, by means of science, the methods of Nature's operation, we shall be able to take her place to perform them for ourselves . . . men will master the forces of Nature; they will become themselves architects of systems, manufacturers of worlds. Man then will be perfect; he will then be a creator; he will therefore be what the vulgar worship as a god.

The reader of this bold and daring manifesto of the environmental ethos of the Enlightenment cannot help but notice how often words such as "conquer," "master," "slave," and "serf" are used to describe the triumph of human genius over the crudities of nature. Western culture is still the prisoner of such views—which, to stress once more, do *not* derive from Christianity, but from the secularism of the Enlightenment, which enthroned humanity as nature's god, lord, and master. Reade argues that religion is little more than the projection of prevailing ideas. "A god's moral disposition, his ideas of right and wrong, are those of the people by whom he is created."

The darker side of things was overlooked amid this uncritical Victorian adulation of science. In 1720, the German philosopher Christian Wolff published a work that laid the foundations for the Enlightenment emphasis on the supremacy of human reason, and the human right to dominate nature. The book had the suitably generous title *Rational Thoughts about God, the Word, the Human Soul and Everything Else*. Its frontispiece is especially revealing of the optimistic view of humanity and nature it proposed. A German country scene is depicted, with mountains, woods, towns, and villages. A glorious sun is dispelling the dark clouds that once overshadowed this pastoral scene. The message is clear: the Enlightenment dispels gloom and darkness and ushers in a new era of light. The dark and gloomy shadows of religion are cast aside as the brilliant light of reason illuminates the world.

Today that frontispiece is seen in a very different way. The license given to humanity by the rationalism of the Enlightenment has led to what seem to be irreversible changes in the atmosphere, leading to ozone depletion and global warming. The sun that now shines down on our rural landscapes becomes deadlier year by year, as the ability of the atmosphere to absorb the force of its ultraviolet radiation is reduced through pollution. The rain that waters the fields is acid, and brings about the death of the woods so carefully depicted in this scene. Far from bringing joy, the Enlightenment has led to the slow death of nature—and perhaps us with it. For among the "taboos" and "superstitions" swept away by the Enlightenment were the ideas that there were limits to human action within the world, and that nature, as God's creation, should be respected and honored. We are now paying the price of that loss.

Religion and the Limits of Exploitation: The Case of Francis Bacon

The importance of religion in constraining human exploitation of nature can be seen from a close reading of the works of Sir Francis Bacon (1561–1626), widely credited as one of the founders of the inductive method so central to modern scientific thought. As is well known, Bacon argued for an inductive approach to truth, which would base its knowledge on the world of experience, rather than innate ideas or arguments.

> There are and can be only two ways of searching into and discovering truth. The one flies from the senses and particulars to the most general axioms: this way is now in fashion. The other derives axioms from the senses and particulars, rising by a gradual and unbroken ascent, so that it arrives at the most general axioms last of all. This is the true way, but as yet untried.

Bacon's approach is thus based on the generation of laws from observation and experiment, by means of an intuitive "analogical leap" from the observed to the unobservable. Bacon did not see this as in any way necessitating, or even implying, that the natural sciences were opposed to religion. As he once commented, "a little philosophy inclineth man's mind to atheism, but depth in philosophy bringeth men's minds about to religion." (A similar comment from the pen of Lord Kelvin, the great nineteenth-century physicist, may be noted here: "I believe that the more thoroughly science is studied, the further does it take us from anything comparable to atheism.")

The importance of religion to Bacon cannot be overstated. Despite his general endorsement of subduing and exploiting nature, Bacon regards religion as setting quite definite limits to what can and ought to take place through human advancement. Bacon was quite clear that there were boundaries imposed upon human knowledge and power over nature by God, and he regularly added qualifying clauses like "as far as God Almighty in his goodness may permit" to his statements concerning the limits of science. Bacon had no doubt that the development of the sciences would change the face of the earth. The new science that he advocated was not simply the passive observation of the world, but the development of means to change its course. It is not enough, he suggested, merely "to exert a gentle guidance over Nature's course"; humanity ought to aim "to conquer and subdue her, to shake her to her foundations." Science was thus to be harnessed to the extension of human empire.

Bacon is well aware of the importance of technology to the advancement of human empire building. The three inventions that Bacon held to be of greatest importance—printing, the magnetic compass, and gunpowder—were all linked with the age of imperial expansion, such as that undertaken by the British under Elizabeth I. As Bacon well knew, the development of the magnetic compass was of critical importance to the great voyages of exploration and subsequently colonization that were undertaken during his time, with gunpowder enabling Britain to enforce its influence globally.

We find this idea explored further in his *New Atlantis,* in which we learn of a research institute by the name of Salomon's House, whose object is set out as follows by its director: "The end of our foundation is the knowledge of causes, and secret motions of things; and the enlarging of the bounds of human empire, to the effecting of all things possible." Even here, however, the notion of divine limitation of human exploitation is clearly assumed, as is evident from the director's final statement: "We have certain hymns and services, which we say daily, of laud and thanks to God for His marvellous works. And forms of prayers, imploring His aid and blessing for the illumination of our labours; and turning them into good and holy uses."

Yet if the idea of God were to be eliminated, no such limits would exist. Humanity would be free to do what it pleased—to set its own limits or to deny them altogether. Back in the seventeenth century, this probably seemed a radical and liberating idea. Surely the abolition of limits was to be equated with the emancipation of humanity? And if God stood in the way of this abolition of limits, then why not eliminate God? It is no accident that deicide lies at the heart of the Enlightenment project of conquering nature. Only when God has been eradicated can humanity do what it pleases.

The Elimination of God

God, in the view of many representatives of the Enlightenment, was responsible for the oppression of humanity. Religion forced people to behave in ways that were at worst outrageous and degrading and at best eccentric. It was *"irrational"*—the greatest and most studied insult that the "Age of Reason" could deliver against its opponents. The existence of God was seen as the last bastion of many traditional beliefs. Eliminate God, and the last remaining obstacle to an unfettered human autonomy would have been removed. People could do what they liked. Remove God, and all things become possible. As Dostoyevsky put it in *The Brothers Karamazov,* "If God is dead, then all things are permitted."

The elimination of God thus became a matter of considerable importance to the restless new generation of early-nineteenth-century Europe. Where French revolutionary armies were bringing about the physical destruction of the church, others felt that a more satisfactory approach lay in eroding the intellectual plausibility of God. In short, the best solution lay in explaining away the idea of God. If this idea could be shown to be a delusion or an invention, it could easily be thrown aside as an irrelevance, or at best a quaint intellectual curiosity. Man would finally be the measure of all things, every conceivable rival having been eliminated. The conquest of nature could proceed without any credible religious barriers being placed in its way. No longer was humanity accountable for its use of nature.

The first great strategy toward this goal was devised by the atheist German philosopher Ludwig Feuerbach (1804–72). Using a highly creative reading of the works of the great German philosopher G. W. F. Hegel, Feuerbach argued that "God" was the product of a sad and lonely human mind. People were anxious about death and longed for meaning in life. So what was more natural than that they should invent the idea of God to console them? Feuerbach's materialism—best seen in his most famous quotation, "man is what he eats" (*Man ist, was er isst*)—led him to draw the conclusion that God was the "objectification" or "projection" of human desires and longings onto some kind of imaginary heavenly screen. What was thrown up was the product of a wounded and lonely individual, seeking consolation and meaning in the cosmos. These ideas were set out in detail in Feuerbach's *Essence of Christianity* (1841) and were seized upon by freethinkers throughout Europe. The work was translated by the English novelist George Eliot and was widely read in an increasingly agnostic England.

The second strategy was developed by Karl Marx (1818–83), who took Feuerbach's argument a stage further. Feuerbach had explained why people believed in God. But the important thing, Marx argued, was to get rid of the idea altogether. This is the basic principle lying behind one of Marx's best-known sayings: "The philosophers have only interpreted the world in various ways; the point is to change it." For Marx, the origins of the spurious human belief in God was socioeconomic alienation. Religion was like opium: it dulled people to the pain and tragedy of life. People believed in God because they needed hope and consolation. Why did they need these? Because they had been alienated by the iniquities and injustices of capitalism. Get rid of human alienation, and the need for religion would disappear with it. Hence the coming of a communist revolution would eliminate the factors that disposed people to believe in God, and ultimately erode faith in God altogether.

A third approach is found in the writings of Sigmund Freud (1856–1939), the founder of psychoanalysis. God was now seen as a wish fulfillment, something that humanity dreamed *about* and hence dreamed *up*. "The interpretation of dreams is the royal road to a knowledge of the unconscious activities of the mind." For Freud, believing in God was a coping mechanism, a way in which people adapted themselves to the harshness of an essentially meaningless world.

This intellectual assault on the notion of God was not without its opponents; indeed, each of the approaches noted above can be subjected to a withering intellectual criticism. Yet it seems as if the cultural mood of the late nineteenth and early twentieth centuries was strongly sympathetic to such ideas and was thus prepared to overlook their weaknesses. As studies of the "reception" of theories make clear, people often accept new ideas on account of their attractiveness, rather than their truth. A theory is born that chimes in with the mood of the time and captures the spirit of a generation. It clothes the mood of a generation in garments of intellectual decency. Western society had grown weary with the idea of God and felt like a change. The deliciously scandalous new theories from Germany offered them the option of precisely such a change.

Was not religion the last shackle limiting the actions and thoughts of an increasingly confident and mature humanity? Did not these approaches point inescapably to the conclusion that, since humanity created God, humanity *was* God? Having dethroned God, was humanity—perhaps for the first time in its history—in a position to ascend to the throne of nature, which previous generations had naively assumed to be occupied by God? No obstacles now remained to the mastery of nature.

This attitude was encouraged by a group of antireligious activists, who saw in the natural sciences a new strategic resource for eliminating religion and enthroning humanity as the lords of the cosmos. So important is this theme that we must explore it in a little detail.

A Temporary Alliance between the Sciences and Atheism

Most historians regard religion as having had a generally benign and constructive relationship with the natural sciences. There were periods of tension and conflict, to be sure—witness the Galileo controversy. Yet on closer examination, these often turn out to have more to do with papal politics, ecclesiastical power struggles, and personality issues than with any fundamental conflict of ideas. As leading historians of science regularly point out, the interaction of science and religion is determined primarily by historical circumstances, and only secondarily by their respective subject matters. There is no universal paradigm for the relation of science and religion, either theoretically or historically. The case of Christian attitudes to evolutionary theory in the late nineteenth century makes this point particularly evident. As the Irish scientist and historian David Livingstone makes clear in a groundbreaking study of the reception of Darwinism in two Presbyterian contexts—Belfast and Princeton—local issues and personalities were of decisive importance in determining the outcome.

In the eighteenth century, a remarkable synergy developed between religion and the sciences in England. Sir Isaac Newton's "celestial mechanics" was widely regarded as at worst consistent with, and at best a glorious confirmation of, the Christian view of God as creator of a harmonious universe. Many members of the Royal Society of London—founded to advance scientific understanding and research—were strongly religious in their outlooks and saw this as enhancing their commitment to scientific advancement.

All this changed in the second half of the nineteenth century. The general tone of the later-nineteenth-century encounter between religion (especially Christianity) and the natural sciences was set by two works: John William Draper's *History of the Conflict between Religion and Science* (1874) and Andrew Dickson White's *History of the Warfare of Science with Theology in Christendom* (1896). Both works reflect a strongly positivist view of history and a determination to settle old scores with organized religion, which contrasts sharply with the much more relaxed and settled relationship between the two that was typical of both North America and Great Britain up to around 1830, reflected in works such as William Paley's *Natural Theology* (1802).

Draper argued that the natural sciences were to be welcomed as the liberators of humanity from the oppression of traditional religious thought and structures, particularly Roman Catholicism. "The history of science is not a mere record of isolated discoveries; it is a narrative of the conflict of two contending powers, the expansive force of the human intellect on one side, and the compression arising from traditionary faith and human interests on the other." Draper was particularly offended by developments within the Roman Catholic Church, which he regarded as pretentious, oppressive, and tyrannical. The rise of science (and especially Darwinian theory) was, for Draper, the most significant means of "endangering her position" and was thus to be encouraged by all means available. Like many polemical works, the work is notable more for the stridency of its assertions than the substance of its arguments; nevertheless, the general tone of its approach would help create a mind-set.

The origins of White's book lie in a lecture he delivered in New York on December 18, 1869, entitled "The Battle-Fields of Science." Here, science was yet again portrayed as a liberator in the quest for academic freedom. The lecture was gradually expanded until it was published in 1876 as *The Warfare of Science*. The crystallization of the "warfare" metaphor in the popular mind was unquestionably catalyzed by White's vigorously polemical writing and the popular reaction to it. The popular late-nineteenth-century interpretation of the Darwinian theory in terms of the "survival of the fittest" also lent weight to the imagery of conflict; was this not how nature itself determined matters? Was not nature itself a spectacular battlefield, on which the war of biological survival was fought? Was it not therefore to be expected that the same battle for survival might take place between religious and scientific worldviews, with the victor sweeping the vanquished from existence, the latter never to appear again in the relentless evolutionary development of human thought and knowledge?

A significant social shift can be discerned behind the emergence of this "conflict" model. From a sociological perspective, scientific knowledge was advocated by particular social groups to advance their own specific goals and interests. There was growing competition between two specific groups within English society in the nineteenth century: the clergy and the scientific professionals. The clergy was widely regarded as an elite at the beginning of the century, with the "scientific parson" a well-established social stereotype. Among these we may number Gilbert White (1720–93), author of the classic *Natural History of Selborne* (1789).

With the appearance of the professional scientist, however, a struggle for supremacy began, to determine who would gain the cultural ascendancy within British culture in the second half of the nineteenth century. The "conflict" model has its origins in the specific conditions of the Victorian era, in which an emerging professional intellectual group sought to displace a group that had hitherto occupied the place of honor. The rise of Darwinian theory appeared to give added scientific justification to this model: it was a struggle for the survival of the intellectually fittest. In the early nineteenth century, the British Association (a professional organization devoted to the advancement of science) had many members who were clergy; by the end of the century, the clergy tended to be portrayed as the enemy of science—and hence of social and intellectual progress. As a result, there was much sympathy for a model of the interaction of the sciences and religion that portrayed religion and its representatives in uncomplimentary and disparaging terms.

The "conflict" model of science and religion thus came to prominence at a time when professional scientists wished to distance themselves from their amateur colleagues, and when changing patterns in academic culture necessitated demonstrating its independence from the church and other bastions of the "establishment." Academic freedom demanded a break with the church; it was a small step toward depicting the church as the (temporary) opponent of learning in the late nineteenth century. But that day and age has passed away. There is no longer any social or cultural need to pitch science and religion against one another. If anything, the twentieth century saw renewed appreciation of their common interests and concerns. Yet there are some powerful voices who seek to keep alive the "conflict of science and religion," subtly changing a specific feature of the academic culture of the second half of the nineteenth century into a permanent aspect of human life and thought.

One such writer is the Oxford zoologist Richard Dawkins, who perpetuates this rather outmoded notion in two ways. First, he continues to use the "warfare" model when the social conditions that lent it credibility have long since disappeared. Dawkins's views would have found a much more positive reception in the nineteenth century. The idea that science and religion are permanently locked in mortal combat is largely sustained by Dawkins's own rather overheated statements, which cause many to overlook the much more pacific, constructive, and responsible attitude of others. As a major professional survey of 1996 made abundantly clear, some 40 percent of working natural scientists hold religious beliefs. Dawkins may well believe that these scientists are profoundly mistaken in doing so and that they find doubtless ingenious ways of explaining away the mismatch between their science and faith. Yet the reality is that at least as many scientists find that their professional duties *support* their beliefs as find that they are in contradiction.

Second, Dawkins continues to portray the natural sciences in bold, uncritically positive terms, failing to note their darker side. A convenient and somewhat puzzling failure to distinguish between "science is about understanding our world" and "science is about the exploitation and degradation of our world" allows him to portray any who express anxieties about the application of the natural sciences as ignorant blockheads who wish we were all back in the Stone Age. But the reality is more complex and disturbing than this allows. Science may once have liberated from superstition; it may now have encouraged us to become enslaved to patterns of consumption and exploitation that can only bode ill for the environment and for ourselves.

This point is made forcefully in Juliet Schor's important work *The Overspent American: Upscaling, Downshifting, and the New Consumer* (1998). Schor's careful survey of contemporary American culture brings out the shocking extent to which America has become preoccupied with patterns of spending, possessing, and discarding, which have ominous implications for the health of the American nation and the global environment. Schor argues the need for a "critical politics of consumption," which will break free from the destructive spiral of the consumerist ethos that currently dominates American culture. You want it? Fine; you can have it. This ethos encourages patterns of consumption that take us far beyond the earth's ecological carrying capacity. The widespread assumption that "the production and consumption of goods have no external effects" is seriously in error and fails to take account of such issues as pollution and depletion of the earth's resources.

It would be unfair to suggest that the sciences are responsible for the rise of consumption-driven agendas. Yet it is essential to note that the *mindset* that affirms that humanity can do what it likes with nature (provided by the Enlightenment) must be linked with an *ability* to tame and exploit nature (provided by science, through technology) before the conquest of nature can proceed. We shall explore this issue further in the following chapter. Science, like religion, has both good and bad points and, like religion, must be challenged when it appears to overstep its limits or encourage inappropriate forms of behavior—and above all, when its proponents declare that they possess "limitless power" to change the world in whatever way they please.

In a remarkable essay entitled "The Limitless Power of Science," the atheist Oxford physical chemist Peter Atkins writes boldly of the failure of all human intellectual and culture endeavors save science: "While poetry titillates and theology obfuscates, science liberates." It is not difficult to see what Atkins has in mind by this assertion. As Freeman Dyson points out in his essay "The Scientist as Rebel," science has often been seen as a liberator—a Prometheus-like figure, bringing freedom from outmoded ways of thought and institutions. Science is a "rebellion against the restrictions imposed by the local prevailing culture." It is a *subversive* activity, a point famously stated in a lecture delivered to the Society of Heretics at Cambridge by the biologist J. B. S. Haldane in February 1923.

History offers us many confirmations of this insight. For the Arab mathematician and astronomer Omar Khayyám, science was a rebellion against the intellectual constraints of Islam; for English physicists of the eighteenth century, it offered a platform for criticizing the pervasive influence of the Church of England; for nineteenth-century Japanese scientists, science was a rebellion against the lingering feudalism of their culture; for the great Indian physicists of the twentieth century, their discipline was a powerful intellectual force directed against the fatalistic ethic of Hinduism. For Primo Levi in the 1930s, the natural sciences were an antidote to the fascism then sweeping through his native Italy: "They were clear and distinct and verifiable at every step, and not a tissue of lies and emptiness like the radio and newspapers."

Yet at the hands of its more uncritical and militant advocates, the natural sciences have ceased to be disciplines that merely offer illuminating, fertile, and important insights into the nature of the world in general and humanity in particular. Science has ceased to be a rebel and has become the establishment against which others now protest, and which excludes its rivals by dismissing them as "irrational," "superstitious," or "religious." The sciences have become conquerors, seeing themselves as possessing "limitless powers" (Atkins) and bent upon occupying territory once held by what are to be regarded as inferior worldviews—such as religion, philosophy, and poetry. Peter Atkins and Richard Dawkins seem to take the view that science will not have completed its task until it has swept away every other area of human reflection, unmasking them as irrational superstitions. Religion is "one of the world's great evils, comparable to the smallpox virus but harder to eradicate" (Dawkins). Dawkins thus assumes a permanent alliance between the sciences and atheism. But is it really that simple?

We are constantly told that we must learn from history. "History repeats itself. It has to. Nobody listens the first time round" (Woody Allen). A cynic might suggest that history is rather less replete with wisdom than some of its more optimistic interpreters might allow. Nevertheless, an important grain of truth nestles within this overstatement. Certain patterns of behavior seem to recur with sufficient frequency to allow them to act as guides to our thinking and acting. Occasionally, they may encourage us; at other times, they must warn us against complacency.

One such lesson is that intellectual systems rise and fall. What seems firmly established today will be discarded tomorrow. "It is known that . . ."

has a disturbing tendency to become "It used to be thought that..." The Marxist ideology which many of my generation saw as the future destiny of humanity, bucking the trend of historical erosion, finally became yet another melancholy example of precisely this same trend. What one generation made the cornerstone of its life and thought is turned on its head by its successors.

There is another lesson of history that must be set alongside the erosion of cultural certainties, giving us cause for reflection and concern. Today's liberator has a disturbing tendency to become tomorrow's oppressor. A worldview that was seen as liberating from the burden of the past itself makes the same demands as that which it overthrew. For instance, Stalin's troops were welcomed as liberators by those who had languished under Nazi occupation in eastern Europe. Too late, these unfortunate people discovered that being liberated by Stalin was a less-than-satisfactory experience. What they had mistakenly termed "liberation" was simply the displacement of one dictator by another.

I have often reflected long and hard on this melancholy state of affairs, perhaps hoping that this pattern would turn out to be nothing more than an illusion. Yet there are few reasons for allowing for this optimistic reading of our past. The same pattern can be seen time and time again. Many in late-eighteenth-century France and early-nineteenth-century England saw atheism as an ideology of liberation, setting humanity free from the burdens of religious duties and subservience to the church. By the end of the twentieth century, atheist states had arguably caused more suffering, death, and oppression than anything they had displaced. The underlying pattern remains the same: the abused turns into an abuser.

So will the sciences, which some saw as liberating, making possible a new era in the history of humanity, eventually come to be viewed as a stealthy oppressor? Atheism was a useful foil with which to combat the excesses of a religious establishment; when it *became* the establishment, atheism showed itself to be intolerant, oppressive, and as murderous as anything it had displaced. Absolute power corrupts absolutely. Perhaps Atkins might like to reflect on the implications of this maxim when he declares, apparently without irony, that the sciences possess "limitless powers." As Mary

Midgley rightly points out in the conclusion to her *Science as Salvation: A Modern Myth and Its Meaning* (1992), the movement that promised to save our world has ended up by bringing it close to destruction: "The party is over. The planet is in deep trouble . . . For the general sanity, we need all the help we can get from our scientists in reaching a more realistic attitude to the physical world we live in." So can the sciences, which arguably have got us into this mess, get us out of it? Modernity saw dominating nature as a virtue. Today it is rightly seen as a vice.

So might postmodernity—widely seen as a cultural reaction *against* modernity—hold the key to the reenchantment of nature?

Postmodernity, Religion, and Nature

The world has undergone radical change in the past fifty years. A fundamental weariness with the pretensions of the Enlightenment has swept through Western culture. There has been a widespread reaction against Marxism, Stalinism, imperialism, colonialism, and many other aspects of the culture of the Enlightenment. "It is the nature of men having escaped one extreme, which by force they were constrained long to endure, to run headlong into the other extreme, forgetting that virtue doth always consist in the mean" (Sir Walter Raleigh). In reacting against the intellectual totalitarian of the Enlightenment, Western culture has set its face against such ideas and turned—often with enthusiasm—to embrace their antithesis.

The Enlightenment's rigid approach to reality, which recognized only one valid way of thinking and representing reality, is now held to be inflexible by postmodern writers. Worse than that; writers such as Michel Foucault argued that the philosophical systems of the Enlightenment encouraged oppression and totalitarianism. Jean-François Lyotard argued that all allegedly "universal" systems, such as Marxism, were totalitarian in their outlook and hence potentially capable of generating mind-sets conducive to "crimes against humanity." If people are convinced of the rightness of their own position, there is inevitably a temptation to control or destroy those who disagree with them. The Enlightenment thus had a hidden agenda: the domination of others and nature. It was inevitable, according to Foucault, that the great totalitarian states of the twentieth century would rise. Only by rejecting the philosophy on which they were ultimately based could humanity avoid future catastrophes of this kind.

Foucault suggests that we look at the way in which totalitarian states behave in order to appreciate the force of his arguments. For example, Stalinism laid down that only Marxism-Leninism would be accepted within the Soviet Union, and deviation from its ideas would not be tolerated. In that Marxism-Leninism was held to be the only true outlook on life, those who disagreed with it were to be viewed as either mentally deficient (and hence required to be placed in mental institutions) or deviationists or anti-revolutionaries (and hence required to be imprisoned, not least to prevent their criminal ideas from spreading within the population). The Soviet prison camps of the Gulag Archipelago were designed both to isolate those who disagreed with the state ideology and to discourage others from following their example.

All this arises, Foucault argues, from rigid authoritarian ways of thinking, which insist that there is only one way of seeing things, and demands that those who refuse to see this are stigmatized as mad, deluded, and evil. It is not difficult to think of such intellectual arrogancies and their negative impact in our own day: the fundamentalist preacher, who regards even the slightest deviation from his beliefs as tantamount to apostasy and consignment to the fires of Hell, or the crudely reductionist scientist, who derides those who fail to share her materialism as at best superstitious idiots and at worst social deviants. Foucault's analysis has an uncomfortable ring of truth about it. It is hardly surprising that a generation fed up with the absolutism of Stalinism (only Marxism tells the truth!), imperialist forms of scientism (only science can tell the truth!), and other such intellectual posturings should turn to the ideas of postmodernity.

Postmodernity, in marked contrast, welcomes moral and intellectual diversity. Where the Enlightenment excoriated religion as irrational and superstitious, postmodernity has strongly advocated the valuing of religion and the recovery of spirituality. Dawkins's crude demand that religion should be eradicated belongs to the totalitarian tradition of the Enlightenment, which fails to take account of the diversity of understandings of the nature of truth and the nature of evidence, and lacks the spirit of tolerance to cope with diversities of belief and practice.

So what is the impact of postmodernity upon our attitudes to the environment? At first sight, the answer might seem self-evident. In that postmodernity represents the rejection of the leading themes of the modern period—including the demand and desire to dominate nature—the rise of the movement can only be good news for the natural world. As we have seen, the secular religion of the Enlightenment is saturated with the demand that humans dominate nature. Positive attitudes to nature would seem to go hand in hand with the rejection of the ethos of the Enlightenment.

In fact, things are not so straightforward. There are any number of ways of viewing nature, and none can be regarded as "right" or "good." They are all to be seen as equally valid. More than that: the idea of "deconstruction," which has gained widespread acceptance within postmodern circles, holds that ideas and values are the outcome of the economic and social circumstances of the community or individual that creates them. The "deconstruction" of ideas, texts, and values involves identifying those hidden factors. There is thus no meaning *in* a text; the text may be interpreted in whatever way the reader chooses.

This process has been applied to nature itself. While rejecting the Enlightenment view of nature as something humanity is to dominate and transform, postmodernity has argued that there are multiple ways of interpreting or approaching nature. The results of this are explored in an important collection of essays entitled *Reinventing Nature: Responses to Postmodern Deconstruction*, which highlights the ecological deficit of these ways of thinking. As Michael Soulé points out in his aptly titled essay "The Social Siege of Nature," things have reached a crisis point:

Living nature—the native species of plants and animals in their natural settings—is under two kinds of siege: one is overt, the other covert. The overt siege is physical; it is carried out by increasing multitudes of human beings equipped and accompanied by bulldozers, chainsaws, plows and livestock. The covert assault is ideological and therefore social: it serves to justify, where useful, the physical assault. A principal tool of the social assault is deconstruction.

As Soulé points out, postmodernism encourages an indefinite number of attitudes toward nature, reflecting the vested interests, social location, and personal affluence of the interpreter. Soulé identifies nine "distinct cognitive formations" concerning nature, including:

- Nature as a mindless force, causing inhumanity inconvenience to humanity, and demanding to be tamed
- Nature as an open-air gymnasium, offering leisure and sports facilities to affluent individuals who want to demonstrate their sporting prowess
- Nature as a wild kingdom, encouraging scuba diving, hiking, and hunting
- Nature as a supply depot—an "aging and reluctant provider" which produces (although with increasing difficulty) minerals, water, food, and other services for humanity

Postmodernity thus does not offer any firm basis for insisting that nature is to be respected, and regarded as something special. In some ways, what is really needed here is an *ontology*—a view of nature that identifies it as intrinsically significant, whether or not we think that it is so. If respect for nature is simply made a matter of human convention or convenience, circumstances will arise when that convention is abandoned or rejected when it ceases to be in the interests of the dominant social grouping. If nature is to be honored, it must intrinsically be worthy of that honor. If it is to be respected, there must be something about it *in and of itself* that justly evokes that respect. Reenchantment depends upon the reaffirmation that nature is special and clarification of how this "special" character is to be understood.

Far from proving the faithful friend of ecology, postmodernity has notably failed to prevent—and, according to Soulé, has actually engendered and legitimated—exploitative attitudes toward nature. Humanity is thought of as a consumer, and nature as that which is consumed. And as Bill McKibben makes clear in his important work *The End of Nature,* the human agenda of controlling reality and dominating nature may have reached a point of no return.

It is entirely possible that Soulé and others have overstated the dangers of postmodern approaches to nature. Nevertheless, support for this concerned attitude can be found in other studies of postmodern culture. An important witness to this is found in a highly significant 2001 article entitled "Why Are Sociologists Naturephobes?," by leading British sociologist Ted Benton. In this paper, Benton points out that his discipline has been responsible for the fostering of an antinature attitude which has highly significant implications for ecology in particular and respecting the integrity of nature in general. Nature is seen by many postmodern writers as placing obstacles in the path of politically correct attitudes and values. For example, why should women be held back on account of some limiting features of female anatomy or physiology? Why should a child who was born with less "natural" talent than someone else suffer any kind of disadvantage as a result? Why should an infertile couple not be able to have children? To sociologists, nature merely erects barriers, creates competition, and behaves in what many of their number see as a reprehensibly right-wing way. Many sociologists are appalled by Darwin's theory of natural selection, which seems to encourage competition and the "survival of the fittest." If this is nature, the sooner it is transcended and eliminated, the better.

Nature *limits* humanity; it is the responsibility of postmodern culture to transcend natural barriers and limits. To convert this into a slogan: *nature limits; technology enables.* It can be seen immediately why this is so serious a matter for anyone concerned with environmental issues. Pristine nature, far from being seen as a glorious inspiration, is a state of brute discrimination on the basis of "natural" qualities such as gender, race, and ability.

This line of argument leads to the view that nature is a flawed starting point, which needs to be developed and modified through technological innovation. Although the language may differ from that of the Enlightenment, the objectives are remarkably similar—to master nature in order to free humanity from the shackles that nature imposes upon us. We want to be liberated from what nature has made us, and become something better. Nature does not define what we can and cannot do; it simply marks the point from which we begin a process of self-improvement. Postmodernity thus values technology just as much as the Enlightenment, seeing it as a tool enabling humanity to change itself and the world. But might that tool, today seen as a liberator, become an oppressor tomorrow? We shall consider this in the following chapter.

CHAPTER FOUR

The Faustian Pact: Technology and the Domination of Nature

 Modern Western culture has chosen to break with its Christian roots and declare that human liberation and fulfillment come about through the domination of the natural world. God and nature are both seen as barriers to unlimited human progress, which can only come about through their subjugation. Yet the modern longing to conquer nature and force it to serve human needs is ultimately nothing but a dream unless the means exist to transform the natural world. Modern technology only came into being when humanity developed the will to transform nature and cast aside any idea that there were limits—whether natural or divine—to what humanity can do.

One of the great themes of modern Western literature is the human longing to dominate nature by mastering its secret forces and powers. Works such as Christopher Marlowe's *Doctor Faustus* and Goethe's *Faust* explore the human fascination with the achievement of power over the elements, whatever the cost. Might not the forces that moved the planets and stars be harnessed and put to human use? Might this not be the key to untold riches, power, and status?

At the technological level, the early modern period witnessed an explosion of creativity, as dreamers, hard-nosed businessmen, and calculating statesmen saw ways of gaining hold on power and hence on fortunes and reputations. In 1620, the influential English philosopher Francis Bacon observed how three previous inventions had reshaped the world as he knew it and had played no small part in propelling England to global prominence during the long reign of Elizabeth I:

> It is well to observe the force and virtue and consequence of inventions, and these are to be seen nowhere more conspicuously than in those three which were unknown to the ancients, and of which the origins, though recent, are obscure and inglorious; namely, printing, gunpowder, and the magnet. For these three have changed the whole face and state of things throughout the world.

As we saw earlier, a fundamental theme of modernism—a term that is usually taken to refer to the cultural mood that began to emerge toward the opening of the twentieth century—is its desire to control and dominate, perhaps seen at its clearest in the Nietzschean theme of the "will to power." Humanity needs only the will to achieve autonomous self-definition. It need not accept what has been given to it, whether in nature or tradition. In principle, all can be mastered and controlled. This desire for liberation was often linked with the mythical figure of Prometheus, who came to be seen as a symbol of liberation in European literature. The rise of technology was seen as paralleling Prometheus's theft of fire from the gods. Defining limits were removed. Prometheus was now unbound, and humanity poised to enter a new era of autonomy and progress. The rise of technology was seen as a tool that would allow humanity to control and shape its environment, without the need to respect natural limitations. It is for this reason that many Enlightenment thinkers were so fascinated by the theme of the "limitless power" of the natural sciences. Might not the sciences enable humanity to break free from its ordained place in nature and become like God, able to change the face of nature?

The writers and thinkers of the Middle Ages were deeply ambivalent over any such developments. Were not these ways of harnessing the secret energies of the universe a violation of the natural order, a rebellion against limits placed upon the power of humanity by God? Many Christian thinkers argued that the narrative of Genesis 3 suggested that the root of human sin and evil was a longing to "be like God"—to have access to secret knowledge and the power that could change the face of nature. Sin represented a refusal on the part of humanity to accept that there was a natural order of things which limited their freedom. Genesis describes the building of the Tower of Babel as an act of human defiance, a longing to share the power and privileges of God. It is a powerful symbol of the human refusal to accept its limits—whether natural or ordained—and to quest for domination and transformation. It is in the birth of this mind-set that the true roots of our ecological crisis lie.

Human Sin and the Degradation of the Environment

If anything can be identified as the enemy of those who care for creation, it is the ruthless human tendency to exploit and refuse to accept that limits have been set for human behavior and activity, either by nature itself or by God. The fundamental element of original sin (as described in Genesis 3) is a desire to "be like God" and to be set free from all the restraints of creatureliness. This resolute refusal to accept a properly constituted place within creation can easily be seen to be linked with the development of tools by which humanity is no longer obligated to operate under any form of moral or physical restraint.

We need to explore in a little more detail the Christian idea of human sinfulness—a notion that the Enlightenment vigorously contested, arguing that it was demeaning to human dignity. Within the Christian tradition, sin can be understood as a refusal to accept limits placed upon humanity on account of its creaturely status. This can be seen from any of the foundational documents of the Christian tradition. For example, the Catechism of the Catholic Church notes that man (the document regrettably uses this noninclusive term) allowed

> his trust in his Creator [to] die in his heart and, abusing his freedom, disobeyed God's command ... In that sin, man *preferred* himself to God and by that very act scorned him. He chose himself over and against God, against the requirements of his creaturely status and therefore against his own good. Created in a state of holiness, man was destined to be fully "divinized" by God in glory. Seduced by the devil, he wanted to "be like God," but "without God, before God, and not in accordance with God."

Human sin perverts the notion of dominion of nature from "care" to "oppression," in much the same way as the pastoral care of an individual can easily be perverted into manipulative control and exploitation. What has the potential for good can be subverted to serve selfish human needs.

Once more, it is essential to ask how Lynn White could manage to overlook this point. Christianity holds that the very essence of sin is human self-centeredness, which causes people to develop skewed relationships with each other, with God, and with the environment. White's analysis at times seems to suggest that Christianity thinks that sin is a good idea, and aims to encourage it. The reality could not be more different. The biblical witness, of central importance for Christians, is emphatic on this point. Human sinfulness leads to environmental damage (for example, see Hosea 4:3; Leviticus 26:16–22). Sin is about a refusal to accept divinely ordained limits and has implications for every aspect of human existence, including the critically important relation to the environment. The Old Testament, for example, stresses the importance of respecting and caring for the land, leaving it fallow as a period of "sabbatical rest."

> For six years you shall sow your land and gather in its yield; but the seventh year you shall let it rest and lie fallow, so that the poor of your people may eat; and what they leave the wild animals may eat.

You shall do the same with your vineyard, and with your olive or-
chard. (Exodus 23:10–11)

The earth is the Lord's, not ours (Psalm 24:1; Leviticus 25:7). We are not
free to do with it what we please, but are accountable and responsible for its
well-being.

The Enlightenment rebelled against what it regarded as arbitrary superstitions regarding the land and rejected the idea of limits being placed upon
humanity in its relation to the earth. Inspired by the myth of Prometheus,
the Enlightenment encouraged the development of new ways—both intellectual and technological—of exploiting the earth and allowing humanity
to break through traditional barriers and obstacles.

The Gospel temptation narrative (Matthew 4:1–11) has featured
prominently in Christian reflections on the potential impact of human sinfulness on the environment. Christ is here depicted as being tempted by Satan. The text invites us to imagine Christ being led by Satan to the top of a
hill, where the kingdoms of the world can be seen, stretching into the distance. Power and authority over all these will be his if only Christ will worship Satan. The implications of Christ's rejection of this power seemed
clear: we are to work within the limits of nature rather than quest for the
ability to change it. Yet technology seemed to make both possible and attractive precisely the powers and possibilities that are here associated with
Satan. The powerful lure of those forbidden powers is perhaps best seen by
exploring the rise of the "Faust" legend and its implications for our theme.

The Faustian Pact and the Limitless Powers of Science

Many were frustrated by the Christian God's apparent embargo on the exploitation of nature and its seemingly limitless powers and resources. Might
there be other more amenable divinities who would permit access to such
powers? Might there be some way of getting around the Christian God, who
seemed to forbid the rape of the environment? This longing for empower-

ment lies behind one of the greatest literary characters of the period—
Dr. Faust, later the subject of major dramatic treatments by Christopher
Marlowe in the sixteenth, Goethe in the nineteenth, and Thomas Mann in
the twentieth centuries. Although the Faust legend is first encountered in its
definitive form in the early sixteenth century, its roots go back much further.
Faust longs to dominate both nature and his fellow human beings. But how
can this be done? How can he gain access to the power that will allow this
to happen?

For Faust, the answer was simple: enter into a pact with the devil. In
the face of Christianity's demand that humanity recognize its limitations,
the Faust myth proposed precisely the opposite. It opened the door to forbidden powers and knowledge that humanity was never meant to possess.
The Faust myth resonated with those who longed for power over nature. If
God would not allow humanity to have this power, there were other deities
that would. Popular Christian piety in the later Middle Ages distrusted the
new technology which was poised to make such an impact on nature and
human history, suspecting that this was spawned through a bargain with the
devil. But what traditional Christianity critiqued and demonized, the Faust
legend affirmed and made culturally respectable.

The idea that humanity should recognize its limits is excoriated and
ridiculed by those committed to a "science as human progress" school. This
is regularly depicted as representing the ludicrous ranting of obscurantists,
locked into outdated ways of thinking and hostile to the idea of progress.
And in many ways, these criticisms are right. Some of the objections raised
against the emerging technology were ludicrous. Yet perhaps they also embodied a grain of truth—a truth so precious that we dare not lose sight of
it. The Faust myth depicts the will to power over nature as a fundamental
human longing that is not sanctioned by God. It can only be achieved by
selling one's soul to the devil. The twentieth century found no shortage of
individuals, corporations, and governments who were more than willing to
sign up to this Faustian pact.

The Faustian defines progress as technological advance, often oblivious
to the moral implications of these developments. In terms of the Faustian
will to power, it is perfectly reasonable to speak of progress—but the

"progress" in question is the amoral progression to better and more efficient ways of achieving power. In this amoral sense of the term, one can
speak of human "progress" from bows and arrows to gunpowder, and from
cannonballs to nuclear missiles. In the end, the human quest for dominance
threatens to destroy our humanity. We unwittingly unleash forces we cannot
control, forces that threaten to overwhelm us ourselves, as well as the natural and human forces we seek to dominate. As Ralph Waldo Emerson put it
in his early "Ode: Inscribed to W. H. Channing":

> *There are two laws discrete,*
> *Not reconciled,*
> *Law for man, and law for things.*
> *The last builds town and fleet;*
> *But it runs wild,*
> *And doth the man unking.*

The Christian understanding of sin is such that it cannot but impact
upon human environmental attitudes. If humanity is radically self-centered,
tension with the environment is inevitable. The tragedy of the situation is
that technology has enabled the secret dark longings of the human heart to
be put into practice. Thoughts of domination and oppression that were
once beyond reach are now entirely possible, through technological advance.
Human sinfulness and technological progress combine to make a lethal
cocktail, in which new and dreadful possibilities are opened up, like Pandora's mythical box—an image to which we may now turn.

Prometheus without Pandora: The Lure of Technology

Many Enlightenment thinkers dreamed that, at last, humanity would be
able to steal the fire of the gods, unrestricted by outmoded beliefs, irrational fears, and superstitious nonsense about the "sacred" character of nature. The rise of the natural sciences led to a growing understanding of
nature, which led directly to new ways of mastering and exploiting nature.

The more humanity could plunder nature, the more it proclaimed its supreme authority over the world. The growth of technology and the ability to overwhelm the natural world was ultimately seen to be linked with human self-esteem. The ability to master nature was seen as the ultimate accolade for human achievement, the crowning glory of the human race.

In older works of literature—such as the Anglo-Saxon epic *Beowulf*—nature was portrayed as a hostile force, threatening to overwhelm humanity. With the rise of technology and the loss of any sense of human limitations, the tables had now been decisively turned. Humanity could conquer the nature that had once held it captive. Humanity had triumphed over both God and nature, proclaiming its mastery of the former by declining to believe in him, and over the latter by forcing it to serve humanity's ends. The Enlightenment was wedded to the idea of human supremacy over all its rivals and set itself the goal of defeating nature and proclaiming the death of God. With these two developments, the triumph of enlightened humanity would be complete.

It is not surprising that many writers of this period turned to the classic myth of Prometheus, finding in this an icon of the autonomy of humanity in the face of the gods. The classic story of Faust—who enters into a pact with the devil in order to gain mysterious powers—is a highly imaginative and influential variation of this legend. According to Hesiod, the gods chose to withhold fire from humanity, seeking to limit their autonomy and potential. Fire was a powerful resource which the gods wished to keep to themselves. In a bold act of defiance, Prometheus wrested fire from the gods and brought it to his people, thus making a divine resource available to humanity.

Whatever Hesiod or Aeschylus may have made of this myth back in classical times, the culture of the Enlightenment saw it as a timeless statement of the human need to break free from arbitrary limitations imposed upon humanity by tradition and especially religion. It is no accident that Mary Shelley provided a subtitle for her famous Gothic novel *Frankenstein* that highlighted the importance of this mythical figure for her era. Her subtitle? *The Modern Prometheus*. As Shelley's novel makes clear, she had serious concerns about where the Enlightenment proclamation of the triumph of science was taking humanity. Yet many chose to overlook her warnings, preferring to stress only the benefits of technology. Prometheus was to be seen only as a liberator, foreshadowing the later liberation of humanity by enlightened reason from the abominable fetters placed upon them by religious tradition and beliefs.

Like what? A simple example is the Old Testament command to allow "the land a sabbath of rest" (Leviticus 25:4). Just as humanity needed a time of rest to recover from its labors, so the land should be accorded the same respect. If humanity treats the land well and respectfully, God "will send the rain in its season, and the ground will yield its crops and the trees of the field their fruit" (Leviticus 26:3). We see here a clear recognition of the link between wise use of the earth's resources and its continued fruitfulness. Yet the obligation to respect the earth in this way is not primarily located pragmatically (although it has pragmatic implications), nor is it grounded in human conventions. It is presented as the will of God, which demands absolute respect. For the Enlightenment, this was primitive nonsense, which could be discarded by modernity as a useless relic of a barbarous past. Why not use the land *all* of the time?

The Greek myth of Prometheus cannot, however, be taken in isolation. Following Prometheus's act of defiance in bestowing the gift of fire on humanity, the gods retaliated. Zeus sent Pandora to Prometheus's brother, Epimetheus, bearing a jar (not a "box," incidentally), which she was instructed not to open. Pandora could not resist the temptation and removed the lid, allowing various calamities to escape and invade the world. Enlightenment writers are generally a little coy about this aspect of the Prometheus myth, which seems to imply that the human attempt to possess the powers of the gods will end with the unleashing of forces and evil that could not have been foreseen, with disastrous results. Mary Shelley's *Frankenstein* is one of the few literary works of this period to express concern about the capacity of human nature to cope with the responsibilities and consequences of the new technology. Sadly, she proved to be a voice crying in the wilderness.

Technological progress has created problems that have led some to wonder whether the advances in question have been entirely positive. In solving old problems (such as once-fatal illnesses) we have created new ones (such as global overpopulation). The entirely laudable goal of understanding how the universe works seems to become linked to a Nietzschean "will to power," thus leading to the mastery, control, and redirection of nature. "The control of nature is a phrase conceived in arrogance, born of the Neanderthal age of biology and philosophy, when it was supposed that nature existed for the convenience of man" (Rachel Carson). We have become the victims of our own success. Our mastery of nature has led to its degradation. Progress has finally been recognized to have its downside.

The Master Race: The Origins of the Autonomous Human

Fyodor Dostoyevsky (1821–81) and Friedrich Nietzsche (1844–1900) both emphasized—though in different ways—the great new insight of the modern era: if God is dead, then everything is permitted. The use to which Nietzsche put this idea presages the worrying situation in which we now find ourselves. Nietzsche's unrestrained assault on reason and traditional morality in the name of human value-creation and will to power has had a major impact on Western culture and is regularly cited as a defining influence on the shaping of twentieth-century culture in the West, and especially in the United States.

Nietzsche developed the idea of Zarathustra—the "Superman"—as a means of challenging the ingrained Christian values of late-nineteenth-century German society. These values, he argued, which lie behind what is generally considered to be good and evil, merely hamper the human potential. For human society to be able to live up to its true potential, we need a new system of values more suited to our ambitions. Nietzsche thus rejects the idea of a transcendent God who gives us changeless norms.

Nietzsche demanded the overthrow of Christian virtues and values, seeing these as merely limiting human potential and autonomy. Christianity held back this new way of being human. In eliminating the idea of God and the values attached to it in his system, Nietzsche is obliged to offer us another godlike figure from whom we may receive our new values, in order to fill the void that is created. In his place, Nietzsche offers us his Superman, an individual who creates values that are firmly rooted in the everyday changing world. This is someone who, by trusting his own intuitive sense of what is good and evil, succeeds better than any other. It is argued that only by following his example can we hope to improve ourselves and our society. This, of course, demands the overthrowing of many traditional values and ideas.

Nietzsche admitted that news of the death of God was a little slow in traveling. "God is dead: but considering the state the species Man is in, there will perhaps be caves, for ages yet, in which his shadow will be shown." Yet the implications of this death were monumental. Without God, human beings were at the center of things, the hitherto uncrowned rulers and unacknowledged legislators who now came into their own. What people proposed, God could not dispose. Many, reading Nietzsche perhaps less critically than they should, saw him as setting out a compelling vision in which humanity was not limited by the way things were or are. They could all be changed. The theme of the will to power was interpreted as a license to change. Did not Nietzsche state that "there are no facts, only interpretations"? And did not this open the way to re-creating and reshaping our mental and physical worlds according to our desires—and open the way to changing the natural as well as the moral world? There was nothing about the natural world that demanded that we leave it as we find it, or accept the limitations it places upon us. We have the power to change it, to re-create it according to our liking.

Nietzsche's words were written at a time when new technological developments were making this a reality. After the Industrial Revolution, machines became capable of transforming nature. Rather than leave nature as we found it, we developed the ability to re-create it, to remaster it to suit our needs and beliefs. The heady combination of Nietzsche's radical philosophy and the rise of a technology that allowed ideas to become practice led to more than the "anthropocentric" view of the world which Lynn White so rightly deplores, yet so wrongly attributes to Christianity. It led to a self-centered humanity, convinced of its right to create its own environment, unimpeded by anything and anyone, and now possessed of the technological ability to turn thought into action. The only canon recognized by technology is that of *possibility*; if something might be done, there is a will to do it—if only to show that it can be done.

To see how this worked out in practice, let us see how ecology fared in the world's most consistently atheistic state—the Soviet Union.

The Master Race in Action: Stalinist Ecology

The Soviet Union was the first nation-state to systematically outlaw belief in God and attempt a programmatic elimination of religion from the public life of the nation. Although this development began under Lenin, it entered a new phase of intensification under Stalin, reaching its zenith in the 1930s. If Christianity is indeed the enemy of ecology—as Lynn White simplistically suggests—one might reasonably expect that the elimination of this faith would lead to a new concern for the environment within the Soviet Union and serve as a model for ecological groups worldwide. In fact, we find a *very* different picture, which provides further disconfirmation of White's crude misrepresentation of the ecological implications of the Christian faith.

We have already seen how the Enlightenment project—which includes Marxism among its many children—is distinguished by its systematic attempt to disenchant nature, changing it from something intrinsically sacred into mere matter for human manipulation. The Enlightenment aimed for the *domination* of a disenchanted and objectified nature, seen simply as something irksome which had to be overcome and mastered for human purposes. There was nothing sacred or special about nature that demanded it be treated with special care and consideration. Stalin's atheism led not to the respect of nature, but to the elimination of any religious restraints on the relentless exploitation of the natural world for political and social goals.

Stalin's Five-Year Plan demanded the restructuring of Soviet society, industry, and agronomy. Official publications stressed that there were "no fortresses that Bolsheviks could not storm" and also stressed the importance of technology in allowing the Soviet Union to put in place its idealized social vision. Any obstacles in the way of this vision would be swept away. The Christian churches, the old ruling classes, and the Western interventionist forces had all been dealt with. What obstacle remained? The answer soon became clear: nature itself.

A number of factors led to a dismissal of environmental concerns during the Soviet Union of the 1930s. The Soviet Union was primarily a preindustrial agrarian farm-based culture, which needed radical transformation if it was to compete for existence in a hostile capitalist world. Everything was ultimately judged on whether it could be bought or sold, and thus generate income for the struggling Soviet economy. A concern for the environment was dismissed as reactionary bourgeois sentimentalism.

Stalin insisted that the Soviet science establishment itself should be reoriented toward a more pragmatic or "utilitarian" view of nature. Stalin had very little use for theoretical science. On the twelfth anniversary of the Bolshevik Revolution, he made it clear that science was subservient to the ideals of Marxism-Leninism:

> All the objections raised by science against the possibility and expediency of organizing great grain factories of forty to fifty thousand hectares have collapsed and crumbled to dust. Practice has refuted the objections of science, and has once again shown that not only has practice to learn from science, but that science also would do well to learn from practice.

Stalin clearly had no interest in examining the potential environmental consequences of such massive grain factories.

As time progressed, Stalin and his acolytes began to view all pure scientists as being at best nuisances and at worst counterrevolutionaries. He suggested that they enjoyed the privileged and protected status that the ecologists had sought to achieve for the few remaining Soviet nature reserves. Stalin increasingly abandoned any attempt to view society and nature as existing in harmony and saw his task as being to conquer and tame nature as if it were a hostile beast. In his masterly study of Soviet ecology at this time, *Models of Nature: Ecology, Conservation and Cultural Revolution in Soviet Russia,* Douglas Weiner remarks:

> Many politically active Soviets viewed nature as an obstacle to socialist construction that had to be conquered. Only a small minority placed equal or greater emphasis on the protection of nature. In the popular literature and the press, antipathy toward harsh nature frequently led authors to anthropomorphize nature. Nature

was portrayed almost as a consciously antisocialist force which needed to be suppressed.

In adopting this attitude, Stalin and his colleagues can be seen as embodying the classic ideas of the Enlightenment. These ideas are set out in Lenin's New Economic Policy (NEP) of 1921, which Stalin developed and applied with remarkable vigor. As Weiner comments:

> The Bolsheviks, as heirs to both the Russian revolutionary intelligentsia and Marxian traditions, had succeeded to important philosophical impulses of the Enlightenment. One was the impulse to desacralize and demystify nature. Desacralization of nature made it ideologically possible for humans to strive to dominate and transform it. Like Bacon, Marx and Engels, NEP Russia had viewed nature coldly, unsentimentally; yet, like those three thinkers, it understood that "nature, in order to be commanded, had to be obeyed." Natural laws could not be altered, only learned and utilized for the benefit of human society, as Lenin and so many others had said. Flowing from this view of nature was strong backing for basic research into the structure of nature; only a fuller understanding of the structure of nature and its laws could permit society to extend its abilities to wrest greater bounties from nature.

Weiner further points out how these attitudes resonated with another theme of the Enlightenment: the inexorable advance of humanity to rule the earth as it pleased. As Pavel Akselrod put it, they were preparing for "a race of gods on earth, of beings endowed with an all-powerful reason and will . . . capable of grasping the world with their thoughts and ruling it." Stalin and his theoreticians saw that nature was an obstacle to this advance, and set out to subdue, master, and dominate it, as a prelude to shaping it to their own ideological ends.

Perhaps lack of access to official Soviet records might explain Lynn White's astonishing blind spot here. White was absolutely right to stress how the imperative to "dominate nature" is central to our present ecologi-

cal crisis. Yet he signally failed to discern its importance within the secular-
ized worldview of the Enlightenment, to which Stalin is heir. White fails to
engage with either Lenin or Stalin in his article, or deal with the obvious
difficulties their ideas pose to his simplistic thesis. White insists that Marxism is merely a "Judeo-Christian heresy," failing to note how it incorporates
the central themes of the Enlightenment precisely where these repudiate the
central themes of the Christian faith.

The strongly hostile attitude to nature that emerged under Stalin at
this time is best seen from the official and semiofficial accounts of the construction of the Baltic–White Sea Canal, which stress how Soviet political
commitment and technological advances enabled the crude forces of nature
to be defeated and put to the service of global Soviet advance. Perhaps the
best-known account of this is found in Maksim Gorky's edited account of
this construction, which depicts the triumph of Soviet ideology and technology over an untamed nature:

> Stalin holds a pencil. Before him lies a map of the region. Deserted
> shores. Remote villages. Virgin soil, covered with boulders.
> Primeval forests. Too much forest as a matter of fact; it covers the
> best soil. And swamps. The swamps are always crawling about,
> making life dull and slovenly. Tillage must be increased. The
> swamps must be drained ... The Karelian Republic wants to enter
> the stage of classless society as a republic of factories and mills.
> And the Karelian Republic will enter classless society by changing
> its own nature.

Gorky attached a motto to his account of the construction of the canal:
"Man, in changing nature, changes himself." The key to the advance of humanity, supremely in the Soviet Union, was the transformation of the untamed natural habitat into something that served the needs of the Soviet
people.

Stalin's antiecological attitudes have much to teach us, and it is essential
that the lessons are learned. First, Stalin was contemptuous of the Russian
pastoralist tradition—the distinguished tradition of thought, well estab-

lished in Russia, and represented by writers such as I. P. Borodin and V. E. Timonov—that held that nature was *intrinsically* special and should be respected for that reason. This view was dismissed as "sentimental" or—even worse, in the eyes of Soviet ideology—"quasi-religious." Stalin and his colleagues drew upon the strongly rationalist and ultimately utilitarian view of nature inherited from the Enlightenment, which desacralized nature. There was nothing "special" about nature. It was there to be used. The intellectual disenchantment of nature was thus identified as the precondition for its despoiling. What you want to rape, you first rob of any sacred character.

Second, the Soviet Union in the 1930s can be seen as the ultimate expression of the self-serving anthropocentric worldview of the Enlightenment. Human self-perfection was the goal; this, however, could only come about by the transformation of *society*, which depended upon the prior transformation of *nature*. Nature was an antisocialist force, which had to be neutralized or eliminated before humans could achieve their historic destiny and actualize their true potential. The effacing of nature was the precondition of human advancement. "Nature was regarded as an enemy to be conquered in the course of the creation of the totally man-made socialist environment; truly, a world without nature as we know it" (Douglas Weiner).

The Rise of Technology

Nietzsche's "will to power" might not have much significant impact on the environment, unless there is some physical means of radically transforming nature at the disposal of his newly liberated humanity. Human beings have always tinkered with their environments, in an effort to increase their quality of life. Lynn White opened his study of the origins of our ecological crisis by reflecting on how such developments took place.

A conversation with Aldous Huxley not infrequently put one at the receiving end of an unforgettable monologue. About a year before his lamented death he was discoursing on a favorite topic:

Man's unnatural treatment of nature and its sad results. To illustrate his point he told how, during the previous summer, he had returned to a little valley in England where he had spent many happy months as a child. Once it had been composed of delightful grassy glades; now it was becoming overgrown with unsightly brush because the rabbits that formerly kept such growth under control had largely succumbed to a disease, myxomatosis, that was deliberately introduced by the local farmers to reduce the rabbits' destruction of crops. Being something of a Philistine, I could be silent no longer, even in the interests of great rhetoric. I interrupted to point out that the rabbit itself had been brought as a domestic animal to England in 1176, presumably to improve the protein diet of the peasantry.

The introduction of the rabbit into England pales into insignificance compared with the vast upheavals and disfigurements of the natural environment caused by the technology of the twentieth century. With the great technological revolutions of the nineteenth and twentieth centuries, humanity gained the ability to change the face of the globe irreversibly. The Enlightenment desire to dominate nature might have expressed itself in trivial and harmless ways if there had been no physical means to manipulate the environment. But with the rise of technology, the dream of the Enlightenment became realizable. Only then did people realize that it might in fact become a nightmare, unleashing forces that could no longer be restrained. The lid was taken off Pandora's fabulous jar, and its contents irreversibly released into the world.

Reviewing the impact of the Enlightenment upon nature, Theodor Adorno and Max Horkheimer noted how its laudable aims appear to have backfired, creating problems that were simply not imagined by those who set it in motion: "Enlightenment has always aimed at liberating men from fear, and establishing their sovereignty. Yet the fully enlightened earth radiates disaster triumphant." These are powerful and deeply evocative words. A movement that aimed at human liberation has ended up trapping humanity in a decaying world. The Enlightenment aimed to achieve human liberation by domination of nature. It has ended up by enslaving people to a dying earth and offering them no alternative home.

One of the most moving reflections on the growth of technology and its impact upon the environment is due to Romano Guardini (1885–1968). In the 1920s, Guardini returned to his native Italy and was deeply distressed at the changes he witnessed as a result of the introduction of technology. He published his reflections in 1926 as *Letters from Lake Como.* The work resonates with a sense of despondency, as its author watches, helpless with despair, as machines obliterate and disfigure what he regarded as the glories of nature. "I saw machines invading the land that had previously been the home of culture ... nature had been given a new shape by us humans."

As he reflected on the changes in his native Italy, Guardini came to the conclusion that the fundamental link between nature and culture had been severed as a result of the rise of the "machine." Humanity was once prepared to regard nature as the expression of a will, intelligence, and design that are "not of our own making." Yet the progress of technology has opened up the possibility of *changing* nature, of making it become something it was not intended to be. Technology offers humanity the ability to impose its own authority upon nature, redirecting it for its own ends. Where once humanity was prepared to contemplate nature, its desire is now to "achieve power so as to bring force to bear on things, a law that can be formulated rationally. Here we have the basis and character of its dominion: arbitrary compulsion devoid of all respect." No longer does humanity have to respect nature; it can dominate and direct it through the rise of technology.

> Materials and forces are harnessed, unleashed, burst open, altered, and directed at will. There is no feeling for what is organically possible or tolerable in any living sense. No sense of natural proportions determines the approach. A rationally constructed and arbitrarily fixed goal reigns supreme. On the basis of a known formula, materials and forces are put into the required condition: ma-

chines. Machines are an iron formula that directs the material to the desired end.

As a Christian philosopher, Guardini was deeply concerned over the failure of technology to respect boundaries. In an earlier phase of human history, technology was used to extend "the range of natural human organs, making possible more acute and accurate seeing and hearing." Guardini here has such instruments as microscopes and telescopes in mind, which allowed nature to be observed and understood more accurately. But technology did not stop there. It moved on from *understanding* nature to *changing* nature— from appreciating the intricacies of nature to exploiting it for human ends.

I find myself resonating with these concerns. As one who loved the natural sciences and was deeply appreciative of the understanding and re- spect for nature that they made possible, I could see the force of Guardini's point. The intellectual *investigation* of nature—which I loved—was being subverted into the *manipulation* and *exploitation* of nature, which I deplored. Most natural scientists will experience this deep sense of unease over the use to which the knowledge of nature is put. Yet that tension is an in- evitable aspect of scientific advance. The otherwise harmless desire to re- shape nature is given a new significance by technology, which extends human power to the performance of all things possible.

Technology thus becomes separated from ethics. The longing *to be able to do things* becomes detached from the question of *whether these things ought to be done.* Many of the physicists working on the Manhattan Project (which pro- duced the world's first atomic bomb) were swept along by their excitement at the new technologies they were developing, and the exploration of as- pects of high-energy physics that up to that point had been impossible. They derived immense intellectual satisfaction from the technical brilliance of their research and the remarkable way in which the first test of an atomic bomb confirmed what had hitherto been theoretical speculation. Yet the fundamental moral question as to whether atomic bombs should be built— or used—seemed to belong to someone else. It was not their problem.

The scientists' interest lay in furthering the human ability to break into new intellectual and technological territory. What was done with these new insights and tools was of no concern to them. In this important case, technology was thus seen as fundamentally amoral, guided by the lodestar of making things possible, irrespective of what this might lead to. Technological advance has lifted humanity to a new moral level—but a level for which technological thinking is simply not equipped to cope. And here lies part of our problem. The Renaissance has often been defined in terms of the "pursuit of eloquence"; the twentieth century can be thought of in terms of the pursuit of technological advance as an end in itself. But for some, technology is just a means to their own ends—and those ends include the pursuit of wealth, power, and influence.

Christianity and Technology

Many scholars have suggested that there is a direct connection between Christianity and the rise of the natural sciences. There is no consensus on this matter. Some aspects of Christianity have nurtured the sciences, others have impeded them. The situation is complex and hence not easy to reduce to simple statements. Nevertheless, there is a wide acceptance that the Christian insistence upon the cosmos as a divine creation leads to the view that the world should be ordered and that humanity should be capable of discerning this ordering. In this sense at least, there are reasons for suggesting a generally positive influence on the emergence and development of the sciences. As Paul Davies points out in *The Mind of God*, "in Renaissance Europe, the justification for what we today call the scientific approach to inquiry was the belief in a rational God whose created order could be discerned from a careful study of nature."

An equally ambivalent role is to be ascribed to Christianity in terms of the evolution of technology. The needs of Christian communities were of no small importance in creating an appetite for technology. At least one scholar has argued that the need for accurate timekeeping was stimulated by the monastic pattern of prayer. If monks were to pray every four hours, they needed some way of keeping track of time. One option was to observe the location of the sun—but, then again, monks were also required to pray at night. And so, to cut a long (and not totally convincing) story short, it is argued that the development of the technology of clockwork was held to be related to the needs of Christian prayer and worship.

The development of clockwork technology is not exactly earthshaking. Clockwork has limited potential when it comes to degrading and violating nature. Yet there is a general principle here, which seems relevant to the bigger issue—namely, that Christians were not opposed to technology when it allowed them to undertake routine or religious tasks with greater efficiency or accuracy. Christians, like everyone else, wanted to be able to travel in greater safety and with greater speed, to be able to heal people from their illnesses and build bigger and safer churches. There was no fundamental religious reaction against the use of technology as such.

In fact, in some areas it was welcomed with open arms. The classic example was the invention of the printing press by Johann Gutenberg in the middle of the fifteenth century. Gutenberg put together a printing system—a way of producing more or less identical copies of a book or pamphlet with unprecedented accuracy and economy. As if to underscore the religious importance of this technological breakthrough, he ensured that one of the first items he printed was the entire text of the Bible.

That the introduction of new technologies could create religious controversy is shown by the debates that accompanied the introduction of chloroform in 1849. Not unreasonably, some suggested that this new technique might be used to ease the pain of childbirth. Some Christians objected, on the grounds that it seemed to counter a text in the book of Genesis: "In sorrow you shall bring forth children" (Genesis 3:16). Other Christians countered by pointing out that the same book, Genesis, spoke of God causing "a deep sleep to fall upon Adam," during which a rib was removed (Genesis 2:21). Did not this offer a religious justification for the use of chloroform? In the event, the debate was settled when prominent social figures in England and America chose to avail themselves of the benefits of chloroform, without apparently provoking divine retaliation. Queen Victoria set the fashion for analgesia after the successful administration of chloroform for the birth of her eighth child, Prince Leopold, in 1853, and two years later for her ninth child, Princess Beatrice. The trend was established earlier in the United States, when in 1847 Fanny Wadsworth Longfellow, wife of the poet, became the first woman in the United States to give birth with the aid of pain relief.

The real issue concerns the use to which technology is put. Christianity has always pointed out the inherent ambivalence of nature. The same human tongue can be used to bless or to curse. The hand that heals can also be the hand that kills. Technology merely extends the capacity of the human agent to do good or evil. In one sense, technology is neutral; the issue concerns the use to which it is put.

What one can use for good, another can use for evil. The same piece of metal can be a sword or a sickle, a spear or a pruning hook (Micah 4:3). Warfarin can act as an anticoagulant to lower the risk of strokes and hence save human lives; it can also act as a powerful poison, capable of wiping out entire populations of rats or humans. The same electric power that sustains a life support machine can also operate a deadly electric chair. The use to which technology is put depends on the interests and agendas of people.

The agenda of the Enlightenment was to dominate nature—physically and intellectually. As the first Soviet Five-Year Plan under Stalin makes clear, the implementation of the radical agenda of human domination of nature was critically dependent upon technology. We find that this attitude even extended to the cosmos in the writings of the Cambridge Marxist crystallographer John Desmond Bernal (1901–71), whose 1929 work *The World, the Flesh and the Devil* set out a vision of science as the means by which humanity could colonize the universe and shape its contours to suit it. This work was subtitled *An Enquiry into the Future of the Three Enemies of the Rational Soul*—the enemies in question being limits fixed on the human capacity to change nature, all of which were being overthrown by scientific progress. Bernal here reflects the Enlightenment ideal, given further impetus through Marxism, of the ability of the human race to mold nature into whatever form suits its purposes.

Noting that the scientific discoveries and technological advances of the last century were "sufficient to revolutionize the whole of human life and to turn the balance definitely for man against the gross natural forces,"

Bernal set out his vision of how this triumphant and all-powerful humanity could now hope to colonize and subjugate other parts of the galaxy:

> Once acclimatized to space-living, it is unlikely that man will stop until he has roamed over and colonized most of the sidereal universe, or that even this will be the end. Man will not ultimately be content to be parasitic on the stars, but will invade them and organize them for his own purposes.

We see here Stalin's attitude to the Soviet landscape being transferred to the entire cosmos. Having screwed up the earth's environment, why not start again somewhere else? The only way of making the environment into a live issue for the anthropocentrism of the Enlightenment is to argue that the survival of the human race is at stake. Yet Bernal and others seem determined to export the same worldview to other stars.

There is no more ecologically fatal combination than easy access to earth-busting technology and commitment to a "let's dominate and transform nature" mentality. Christianity is loosely implicated in both these developments; it is not, however, the prime suspect. For that, we need to look to the anthropocentrism of the Enlightenment and its baleful legacy, which reached its zenith under Stalin.

Part of that rationalist legacy is an understanding of nature that erodes it of any mystic and sacral character. To understand how this happened, and why it is of such importance to our theme, we need to explore the rise of the mechanical universe and the outlook on life that this sustained.

CHAPTER FIVE

The Mechanical Universe and the Desacralization of Nature

One of the most significant achievements of modern Western civilization has been to rob nature of any mystic or sacral qualities and to represent it as a resource rightfully at the disposal of humanity. This understanding perhaps reached its zenith in the West in the decades immediately following World War II. Many have laid the blame for this firmly at the feet of an uncritical scientific positivism, which saw intellectual mastery of nature as the prelude to its economic exploitation. To plunder and exploit nature on the massive scales now possible through technological innovation, it was first necessary to get rid of any obstacles to this process of wholesale pillage—such as the idea that nature was itself sacred or that humanity was under an obligation to tend the natural order.

In a remarkable essay entitled "The Empty Universe," the English literary critic and theologian C. S. Lewis showed how the disenchantment of nature led to its being viewed as nothing but the projection of human ideals and longings. There was nothing special about nature, save our subjective perceptions of how it was to be viewed.

> At the outset the universe appears packed with will, intelligence, life and positive qualities; every tree is a nymph, and every planet a god. Man himself is akin to the gods. The advance of knowledge gradually empties this rich and genial universe, first of its gods, then of its colours, smells, sounds and tastes, finally of solidity itself as solidity was originally imagined. As these items are taken from the world, they are transferred to the subjective side of the account; classified as our sensations, thoughts, images and emotions.

Some may *feel* that nature is sacred or enchanted; it is not so in reality.

Science is an intellectual joy, allowing us to grasp and appreciate the complex beauty of the natural order. Yet too often, this becomes the pretext for domination and exploitation rather than an invitation to respect and honor. The noted physicist Robert Boyle spoke of the scientist as a "priest in the temple of nature," evoking the rhetoric and imagery of respect and reverence. Yet things have changed. The dominant theme of our times is that nature has become *disenchanted*—robbed and emptied of whatever mystery and sanctity it once was believed to possess. Those who adopt intellectually aggressive approaches to the natural sciences demand that we bring our ways of thinking about the world into line with their theories, and insist that they be allowed to police what may and may not be believed about the world and its purposes. The model of nature that has been implicated in this process of disenchantment is that of the universe as a mechanism— as a machine, devoid of purpose or goals. In this chapter, we shall consider how this state of affairs came about and what can be done about it.

In recent years, growing attention has been focused on the "intellectual slaying of nature" in the early modern period. An increasing number of writers have argued that a fundamental shift in attitudes to nature can be discerned in the writings of Galileo Galilei, Isaac Newton, Francis Bacon, and René Descartes. The outcome of this new stance toward nature was, in the view of many writers, inevitable. Nature was portrayed as something just waiting to be dominated, remolded, and refashioned to suit the needs of humanity.

This kind of approach has provoked something of a backlash from mainline scientific writers, who have seen it as yet another instance of the growing chorus of irrational and antiscientific attitudes in Western society. The ideas set out in books such as Carolyn Merchant's *Death of Nature* (1980) are held to be nothing less than a crude frontal assault on the scientific method and all the benefits it has brought humanity. Yet the points Merchant and others make cannot be ignored. Whether we like it or not, the Western scientific tradition includes the nourishing of some deeply disturbing attitudes toward nature, which many argue underlie the pillage of nature in our own era.

The origins of this deplorable trend are generally held to lie in the writings of Sir Francis Bacon, who is regularly accused of the use of rape and torture imagery in his writings that relate to the natural world—such as his *New Atlantis,* published shortly after his death. The particular criticisms directed against Bacon rest on his idea of "hounding" nature in order to identify the particulars on which the natural sciences are based. Yet it is questionable whether Bacon intended these images to be read in quite the way some of his critics have suggested. The accusation that Bacon encouraged his readers to think of "raping" or "torturing" nature is difficult to sustain. A careful study of his writings suggests that it is perhaps easier to read such ideas into his writings than to find them explicitly stated. Thus Bacon's suggestion that "you have to but hound nature in her wanderings, and you will be able when you like to lead and drive her afterwards to the same place again" cannot reasonably be taken as deploying torture or rape imagery. It is simply Bacon's description of controlled experiments aimed at making results replicable. Those ideas of torturing and raping nature, however, were not slow to emerge in the writings of his more uncritical followers, some of whom attributed them to Bacon himself.

Now, it is entirely understandable that some within the scientific community have reacted with scarcely concealed anger against any criticisms directed against the scientific project. At times, that irritation is entirely justified, not least when serious misrepresentations of the methods and assumptions of the sciences are involved. Yet it is important to appreciate that there are many who are respectful of the natural sciences, and knowledgeable of their methods, who still wish to urge caution and express concern over some of the assumptions that have come to be associated—rightly or wrongly—with the scientific project and the perhaps unintended consequences of its conception and application. To suggest that one of the most unfortunate and unintentional outcomes of the scientific project has been the disenchantment of nature is not to fall victim to some fashionable irrational or antiscientific polemic. It is to express concern about what seems to be a link between a growing disrespect for nature on the part of humanity and a way of *conceptualizing* nature which seems to underlie this attitude. The way we *see* the world shapes the way we *treat* that world.

In this chapter, we shall explore how modern understandings of nature emerged, displacing older models which, though in some ways unsophisticated, nevertheless embodied critically important insights concerning the relation of humanity to its environment. To begin with, however, we need to explore why it is necessary to use "models of nature" in our engagement with the environment.

Why "Models" of Nature?

Why do we speak of "models of nature" and not just simply speak of "nature"? The basic issue here is the irreducible complexity of the idea of nature. As Kate Soper points out in her important study *What Is Nature?*, the word "nature" is one of the most complex terms in the English language. She argues that there are three major general ways of using the word "nature" in normal human discussion.

1. Used as a realist concept, "nature" refers to the structures, processes, and casual powers that are at work within the physical world, and are investigated by the natural sciences.

2. Used as a metaphysical concept, "nature" denotes a category that allows humanity to affirm its distinctive nature and identity in relation to the nonhuman. Humanity occupies some special place that sets it apart from the rest of the natural order. The term

"culture" is often used to refer to the distinctive sphere of human activities.

3.   Used as a "surface" concept, the term refers to ordinarily observable features of the world. This is perhaps the most widely used sense of the term in modern ecological discourse, in which a contrast is often drawn between nature and an urban or industrial environment, often to highlight how nature has been violated, and thus to emphasize the need for conservation and preservation of the natural habitats that remain.

These three senses of the term interact with each other. Their distinctive ideas spill over into their respective ways of thinking, making it impossible to separate them neatly into some kind of hermetically sealed capsules.

Yet the difficulties in making sense of nature have only begun. Nature is approached from different angles and viewed through different prisms. The importance of this point has been stressed by philosopher of science N. R. Hanson, who pointed out that we do not simply "see" things; we "see" things *as* something. There is a covert process of interpretation implicit within the process of observation. We observe nature through a filter, a set of assumptions, which conditions what we think we are seeing.

Hanson illustrates this important point in a number of ways. For example, he asks us to imagine two different categories of people observing the dawn. One holds a Copernican view of the solar system, in which the earth and other planets are understood to orbit the sun, which stands (more or less) at the center of the system. The other is more antiquated—perhaps a medieval writer transported in time, or one of the 30 percent of the readers of the French newspaper *Le Figaro* who, in response to a recent survey, recorded their belief that the sun goes around the earth. (An entirely natural view for the French, their critics unkindly suggest.) One observer thus thinks of—or *sees*—the earth moving around the sun, and the other the sun moving around the earth.

Now imagine them both watching the dawn on a cold and frosty winter morning. What do they *see?* In one sense, they see the same thing. Yet in another sense, they see something quite different. One sees the sun climbing above the horizon as a result of the sun's movement westward; the other sees the earth's rotation to the east as causing the sun to *appear* to move westward. The horizon falls away; the sun does not really "rise."

This is a very simple example, illustrating a complex point. Given that the idea of "nature" is a complex notion, there are a number of ways of zooming in on the idea, which make it more manageable by focusing only on certain of its aspects. It is an approach widely used in the natural and human sciences that has proved its use over many years. Each model illuminates part of the totality, offering a specific angle on a more complex whole, allowing us to build up a greater overall understanding.

We find these models used widely in the natural sciences, social sciences, and Christian theology—to mention only those disciplines that have a direct bearing on the topic of this book. An example may help us to appreciate the point at issue. In December 1910, the noted physicist Ernest Rutherford developed a very simple model of the atom, using the solar system as an analogue. An atom consists of a central body (the nucleus), in which practically the entire mass of the atom is concentrated. Electrons orbit this nucleus, in much the same way as the planets orbit the sun. Whereas the orbits of the planets were determined by the gravitational attraction of the sun, Rutherford argued that the orbits of the electrons were determined by the electrostatic attraction between the negatively charged electrons and the positively charged nucleus. The model was visually simple and easy to understand, and offered a theoretical framework that explained at least some of the known behavior of atoms at this time.

A further well-known example of the use of models in the natural sciences dates from the 1920s, when the distinguished Danish physicist Niels Bohr found that the behavior of light was so complex that it was best visualized using two irreconcilable models. At times, light was best conceived as a wave, at times as a particle. Bohr was quite clear that he was making no fundamental statement concerning the *nature* of light; he was simply offering visual models or analogies that helped account for its behavior.

So what's the problem? Why should this well-established practice, which is of proven value, cause anyone any problems? The answer to this question is immensely important and underlies the flawed approaches to nature that have dominated the modern period. There are two fatal concep-tual errors that can be made in relation to models.

1. Failing to realize that these are only models—that is, analogies or metaphors that are intended to help us visualize what is often an abstract entity, and understand at least one aspect of its behavior by comparison with a known system. To say that "A is a model for B" is not to say that "A *is* B," but that "A is *like* B in certain re-spects." Thus the atom is *not* a miniature solar system; the Ruther-ford model merely points out that we can understand some important aspects of the behavior of atoms if we think of them in this way. The model offers a visualizable representation of a sys-tem, which assists explanation and interpretation. Models are to be taken *seriously* (in that they clearly bear some relation to the system that is being modeled); they are not, however, to be taken *literally*.

2. Insisting that only *one* of a series of models is required to offer a total description of the complex entity in question. In effect, this is a form of reductionism, demanding that something complex and multifaceted must be reduced to only one of its many aspects. This often reflects sheer intellectual laziness, an unwillingness to engage with something that makes demands of it. More worry-ingly, it is the preferred approach of those with reductionist agen-das, determined to eliminate aspects of reality they find uncomfortable. The simple truth of the matter is that the more complex a system, the more different models are needed to de-scribe it—and hence the greater the distortion and degradation that results through the wooden insistence that only *one* of these may be used.

"Nature" is the most complex entity conceivable, and it is therefore to be expected that a wide range of models has been used to help explain and interpret it. These include:

- a living organism
- a book
- a mirror
- a theater
- a woman
- a clockwork mechanism

Each of these illuminates some aspect of the immensely complex and important aggregate of ideas that we know as nature. To illustrate this, we will look at one model of nature, which enjoyed considerable popularity at the time of the Renaissance. Galileo Galilei (1564–1642) thought of nature as a "grand book," which could be read by those who were expert in the language in which its text was written—namely, the language of mathematics:

> Philosophy is written in this grand book, the universe, which stands continually open to our gaze. But the book cannot be understood unless one first learns to comprehend the language and read the letters in which it is composed. It is written in the language of mathematics, and its characters are triangles, circles, and other geometric figures without which it is humanly impossible to understand a single word of it.

This basic framework is of considerable importance in relation to the development of the "two books" tradition within Christian theology, especially in England, which regarded nature and Scripture as two complementary sources of our knowledge of God. Thus Francis Bacon commended the study of the "book of God's word" and the "book of God's works" in his *Advancement of Learning* (1605). This work had considerable impact on English thinking on the relation between science and religion. In his 1674 tract *The Excellency of Theology Compared with Natural Theology*, Robert Boyle noted that "as the two great books, of nature and of scripture, have the same author, so the study of the latter does not at all hinder an inquisitive man's delight in the study of the former." At times Boyle referred to the world as "God's epistle written to mankind." Similar thoughts can be found expressed in Sir Thomas Browne's 1643 classic *Religio Medici:*

> There are two books from whence I collect my divinity. Besides that written one of God, another of his servant, nature, that universal and publick manuscript, that lies expansed unto the eyes of all. Those that never saw him in the one have discovered him in the other.

This metaphor of the "two books" with the one divine author was of considerable importance in holding together Christian theology and piety and the emerging interest and knowledge of the natural world in the seventeenth and early eighteenth centuries. There can be no doubt that it offered a major theological motivation and incentive to the committed investigation of nature by Christian natural philosophers, who were completely persuaded that the study of the works of God led to a glimpse of the mind of God.

This metaphor or model of nature as a book casts light on a number of facets of the natural world and has the potential to stimulate a productive and creative encounter with it. Yet our concern has more to do with ways in which this model of nature—or any other single model—can be abused, primarily in two ways:

1. By failing to appreciate the analogical or metaphorical status of the language. It is not being suggested that nature *is* a book; merely that the image of a book is helpful as a lens or prism through which to view nature. To use Hanson's language, *seeing* nature *as* a book allows it to be "read" in potentially helpful and creative ways.
2. By insisting that only this model, *and no other,* may be used in determining the character of nature. Because we choose to use the model of nature as a book, all other models are declared to be improper, illegitimate, or redundant. This is most emphatically not the case. Such models are complementary, not exclusive; the more

complex the system, the greater the number of models required to illuminate and explain its various aspects.

With this point in mind, we may turn to consider the rise of the mechanical worldview—the approach to nature that declared that nature was a mechanism, like clockwork, and that this model rendered all other models redundant and irrelevant. The implications of this transition were disastrous, and it is important to consider them fully.

The Clockwork Universe: Nature as a Mechanism

In an important study of the scientific revolution of the seventeenth century, Richard S. Westfall demonstrated that the concept of nature that now came to predominate had four distinct features:

1. *Quantification.*

One of the most important achievements of the period was the demonstration that the patterns of natural behavior—such as the motions of the planets or the falling of objects—could be described mathematically. Alexandre Koyré's phrase the "geometrization of nature" serves as an admirable summary of this new attitude to nature. Both Kepler and Galileo argued for the ability of geometry to render the character of the universe. For Galileo, the "book of the universe" was "written in the language of mathematics, and its characters are triangles, circles and other geometric figures."

2. *Mechanization.*

Descartes and other writers of the period abandoned any notion of nature as an organic entity and compared it to a mechanism, such as a clock (an image popularized by Robert Boyle). Nature was a world of passive matter, made up of individual particles or atoms, whose behavior was governed by mechanical laws. Even human beings could be thought of as machines. Giovanni Alfonso Borelli offered an account of the human skeletal and muscular system that treated it as a system of levers and applied forces. Archibald Pitcairne suggested that the human circulatory system was simply a type of hydraulic machine. Christiaan Huygens argued that the universe was constructed of one common matter, which expressed itself in different sizes, shapes, and motions.

3. Nature as the "other."

Part of Descartes's intellectual program was the development of a conception of nature as the "other," through a fundamental challenge to any suggestion of a genuine affinity between the natural order and its human observer. As the century developed, the "otherness" of nature was reinforced primarily through a growing awareness of the size of the universe. In 1698, Francis Roberts suggested that the light from the nearest star would take at least as long to reach the earth as a ship would take to reach the West Indies—namely, six weeks. As the speed of light was then reckoned to be about 48,000 leagues per second, it is clear that the age possessed at least some apprehension of the vastness of the universe.

4. Secularization.

Here, Westfall draws attention to a growing trend to accept the authority of experimental observations, and reflection upon them, in dealing with questions of science, rather than in turning to religious sources of authority. Given the misleading associations of the term "secularization," Westfall's point is probably better expressed in terms of a growing emphasis on the autonomy of nature. Nature was to be examined and explained on its own terms. It was not divine, and was not given any special status or privileges in the face of human inquiry and advance. An older view of nature, which held that it possessed a position of privilege and dignity and held humans accountable for how they used it, was swept aside. Secularization eliminated both any special divine status of nature and any human responsibility toward it. Henceforth, human attitudes to nature would be defined on utilitarian grounds: exploit nature while you can; when your own existence is threatened by its degradation, start treating it with greater respect. Humanity has become the measure of all things.

The overall impact of this new understanding of nature was to eliminate any sense that humanity and nature belonged together or that their destinies were interlocked. Nature was the "other," something to be quantified as one might count coins, weigh out grain, or measure the distance between towns. It was not something to which humans could *relate*. To use the language introduced by the Jewish philosopher Martin Buber in the 1920s, nature was to be seen as an "It" rather than a "Thou," something that is *known about* rather than *known*. Nature is a thing, and is "living" only in the sense that a machine generates activity. And the model that seemed to sum up this worldview was to see nature as a clockwork mechanism.

The invention of clockwork fascinated generations and opened the way to new technological advances. Perhaps the most famous of these was the invention of a highly accurate chronometer, which allowed navigators to calculate their longitude with unprecedented precision—a tale elegantly told in *Longitude*, Dava Sobel's account of how Thomas Harrison conquered what was perhaps the greatest technological problem of his age—the calculation of longitude. The precision and reliability of clockwork seemed to many to point to machines as the best models for nature. Older ways of thinking about nature—such as nature as a living organism—were abandoned, as a new way of conceiving nature swept its rivals aside.

The rapid advances in astronomy played no small part in this shift. Fresh from his triumph in showing that the planet Mars orbited the sun elliptically, Johann Kepler (1571–1630) made it clear that his discovery had convinced him that the universe was to be conceived as a machine: "I am much occupied with the investigation of the physical causes. My aim in this is to show that the celestial machine is to be likened to clockwork, rather than to a divine organism."

The growing tendency to think of nature as a vast machine was given a powerful new stimulus through the work of Sir Isaac Newton. Newton's great achievement was to show that just about every aspect of planetary motion could be accounted for by the laws of mechanics. The same universal principles that governed the falling of an apple to the earth applied also to the orbits of the planets and their moons. In effect, Newton treated the planets as falling bodies that never quite managed to land on the sun, and as a result merely orbited it. Admittedly, there were one or two problems. Newton's calculations suggested that the planetary orbits were intrinsically unstable, so that the planets would eventually crash into the sun. Yet the explanatory power of this mechanical model of the solar system was so compelling that such minor irritations could be overlooked. The universe, it seemed to many, was like a giant and intricate clockwork mechanism. To some, the universe was not just *like* a machine; it *was* a machine. The model had become the reality.

The success of the mechanical model was compelling and had important religious implications. For Newton and his contemporaries, the intricate mechanisms of the universe represented a clear indication of divine design and construction. The English theologian Richard Bentley (1662–1742) and others developed a series of arguments for the existence of God, which drew upon the explanatory successes of Newton's clockwork universe. How could anyone doubt the existence of such a God when the universe had so evidently been designed?

This approach reached its zenith in the early Victorian period, especially in the writings of William Paley, whose *Natural Theology; or Evidences of the Existence and Attributes of the Deity, Collected from the Appearances of Nature* (1802) had a profound influence on popular English religious thought in the first half of the nineteenth century, and is known to have been read by Charles Darwin. Paley was deeply impressed by Newton's discovery of the regularity of nature, especially in relation to celestial mechanics. It was clear that the entire universe could be thought of as a complex mechanism, operating according to regular and understandable principles.

For some deist writers, this suggested that God was no longer necessary. A mechanism could operate perfectly well without the need for its creator to be present all the time. One of Paley's more significant achievements was the rehabilitation of the idea of the "world as a mechanism" within a Christian perspective. Paley managed to transform the clockwork metaphor from an image associated with skepticism and atheism to one associated with a clear affirmation of the existence of God.

For Paley, the Newtonian image of the world as a mechanism immediately suggested the metaphor of a clock or watch, raising the question of

who constructed the intricate mechanism that was so evidently displayed in
the functioning of the world. One of Paley's most significant arguments is
that mechanism implies "contrivance." Writing against the backdrop of the
emerging Industrial Revolution, Paley sought to exploit the apologetic potential of the growing interest in machinery—such as "watches, telescopes,
stocking-mills, and steam engines"—within England's literate classes.

The general lines of Paley's approach are well known, not least through
the writings of his critics, such as Richard Dawkins. Indeed, Dawkins's important work *The Blind Watchmaker* (1986) is to be seen as a critique specifically of the ideas associated with Paley. During Paley's time, England was
experiencing the Industrial Revolution, in which machinery was coming to
play an increasingly important role in industry. Paley argues that only someone who is mad would suggest that such complex mechanical technology
came into being by purposeless chance. Mechanism presupposes contrivance—that is to say, a sense of purpose and an ability to design and fab-
ricate. Both the human body in particular and the world in general could be
seen as mechanisms that had been designed and constructed in such a manner as to achieve harmony of both means and ends. It must be stressed that
Paley is not suggesting that there exists an analogy between human mechanical devices and nature. The force of his argument rests on an identity: nature *is* a mechanism and hence was intelligently designed. He makes this
point by comparing a watch with a stone:

> In crossing a heath, suppose I pitched my foot against a *stone*, and
> were asked how the stone came to be there. I might possibly an-
> swer, that for any thing I knew to the contrary it had lain there for
> ever; nor would it, perhaps, be very easy to show the absurdity of
> this answer. But suppose I had found a *watch* upon the ground, and
> it should be inquired how the watch happened to be in that place. I
> should hardly think of the answer which I had before given, that
> for any thing I knew the watch might have always been there. Yet
> why should this answer not serve for the watch as well as for the
> stone; why is it not admissible in the second case as in the first? For
> this reason, and for no other, namely, that when we come to

inspect the watch, we perceive—what we could not discover in the stone—that its several parts are framed and put together for a purpose, e.g., that they are so formed and adjusted as to produce motion, and that motion so regulated as to point out the hour of the day; that if the different parts had been differently shaped from what they are, or placed after any other manner or in any other order than that in which they are placed, either no motion at all would have been carried on in the machine, or none which would have answered the use that is now served by it.

Paley then offers a detailed description of the watch, noting in particular its container, coiled cylindrical spring, many interlocking wheels, and glass face. Having carried his readers along with this careful analysis, Paley turns to draw his critically important conclusion—that "the watch must have had a maker—that there must have existed, at some time and at some place or other, an artificer or artificers who formed it for the purpose which we find it actually to answer, who comprehended its construction and designed its use."

Now, Paley is not singing the praises of the Industrial Revolution, which devastated England's rural economy through encouraging migration to the cities, led to the degrading and disfiguring of landscapes through the intensive and unregulated mining of coal and iron ore, and brought about new and inhumane working conditions. Much of Paley's career was spent working as a clergyman in the English diocese of Carlisle, the home of England's Lake District. Paley's concern is not to endorse the new technology that was transforming England, physically and socially. He was attempting to show how the new fascination with machinery and technology, which many had concluded led *away* from faith in God, could actually be interpreted to lead *to* faith.

Yet Paley was in a minority. Most had drawn the conclusion that the clockwork God of the Newtonian worldview was an irrelevance to life. God may have wound up the cosmic clock, but it was now happily ticking away without any need for divine oversight. God became rather like a retired constitutional monarch—a figurehead, with no real involvement in the day-today running of the universe. If he were to cease to exist, nobody would notice. The universe would continue to tick without divine assistance.

What had once seemed to be a promising alliance between science and religion thus led to a growing estrangement. The Newtonian system suggested that the world was a self-sustaining mechanism that had no need for divine governance or sustenance for its day-to-day operation. By the end of the eighteenth century, it seemed to many that Newton's system actually led to atheism or agnosticism rather than to faith. This can be seen reflected both in Pierre-Simon Laplace's *Treatise of Celestial Mechanics* (1799), which effectively eliminated the need for God (either as an explanatory hypothesis or as active sustainer) in cosmology, and in the writings of the poet William Blake (1757–1827), in which the Newtonian worldview is at times equated with Satan.

So what's the problem? Why did the model of nature as a machine cause so many people such concern? Three answers may be given.

1. This one "model" of nature came to be seen as the *only* valid way of conceiving the universe. It was seen as displacing all other models of nature, which were now regarded as obsolete.
2. Some of those models that were unnecessarily and improperly declared to be outmoded by the new mechanical model were strongly conducive to positive and responsible human attitudes to the environment.
3. The mechanical universe encouraged the emergence of the "ghost in the machine"—a radical dualism of mind and body which encouraged the idea that humanity was utterly distinct from nature.

We shall consider each of these points in more detail in what follows.

The Mechanism as the Only Valid Model of Nature

The idea that nature was simply a giant and particularly complicated machine began to be explored in the sixteenth century. Works such as Agostino Ramelli's *Various Ingenious Machines* (1588) and Jacques Besson's *Theater of Machines* (1569) propagated the notion that the "age of the machine" had arrived and was here to stay. Yet the idea that nature was simply a mechanism was resisted by many at the time. The great Renaissance scientist and artist Leonardo da Vinci (1452–1519) may have enjoyed designing flying machines (even if they were doomed to remain obstinately on the ground), but he had no doubt that nature was rather more than a machine: "We can say that the earth has a vegetative soul, and that its flesh is the land, its bones are the structures of the rocks, its blood is the pools of water, and its breathing and its pulse are the ebb and flow of the sea." For da Vinci, nature was to be thought of—at least in some of its aspects—as a living organism. Da Vinci was perhaps one of the most mechanically minded thinkers of his generation; yet he saw no difficulty in thinking of nature in organic terms. The triumph of the mechanical model of nature in the following century meant that such important complementary insights were discarded.

We are often right in what we affirm and wrong in what we deny. There is no doubt that the new awareness of the regularity of the universe pointed to the appropriateness of using the machine as a metaphor for nature. The orbits of the planets were as regular as clockwork; why not use a clockwork machine as a model for the solar system? The logic seemed unassailable. Yet nature is an immensely complex entity, which cannot be reduced to even a few perspectives, let alone a single one. To do justice to the richness and diversity of nature, it is necessary to deploy a range of models, analogies, and metaphors. This insight seems to have been overlooked, however, by the headlong rush of well-meaning scientists and philosophers to embrace the mechanical worldview during the eighteenth century. Their legacy still remains a powerful component of modern Western culture.

The remarkable successes of the Newtonian worldview led to the triumph of a strongly reductionist view of nature. Where wiser voices argued that a clockwork machine was only one analogy of nature among many, others declared that its considerable explanatory successes demanded that it be recognized as the *only* valid model of the universe. Nature was not just like a clockwork machine in some respects; it *was* a machine. As the full extent of the triumph of the human mind in discerning the ordering of the world became clear in the seventeenth and early eighteenth centuries, the credibility of the mechanical model of nature soared, to the detriment of its nonmechanical rivals. To affirm that nature could be thought of as a mechanism was held to deny that it could be thought of as anything else.

The mechanical approach was now extended to virtually every aspect of the natural world. Newton seemed to be perfectly content to limit his mechanical model of nature to the astronomical world of planets and moons; his successors sensed that new triumphs of explanation lay over the horizon, and pursued the mechanical worldview still further and more radically. The virtually simultaneous invention of "adding machines" by Blaise Pascal in France and John Napier in Scotland led Thomas Hobbes (1588–1679) to argue that the human brain was simply a calculating machine: "When a man reasoneth, he does nothing else but conceive a sum total . . . Reason is nothing but reckoning, that is, adding and subtracting." Where Newton's followers saw the clock as a model for the solar system, Hobbes saw the new adding machines as models for the human brain.

By the end of the eighteenth century, the new "mechanical philosophy of nature" had triumphed. There were still pockets of resistance to, and occasionally—as in the case of Romanticism—outright rebellion against, the mechanical philosophy. William Blake wrote scathingly of "Bacon, and Newton, sheath'd in dismal steel"—a critical allusion to the cold ideology of mechanism. There can be no doubt, though, that the notion of nature as a machine had captured the popular imagination. Even William Paley's appeal to the mechanism of the watch as evidence for the existence of God presupposed precisely the mechanical philosophy he wished to displace.

All other ways of thinking about or looking at nature were now ridiculed. The main consequence of this new understanding of nature was the elimination of any sense of nature as a living entity, with humanity as an integral aspect of the natural order. Nature was defined over and against humanity, which was physically and intellectually *distanced* from the natural order. Nature is the "other," to be conquered and subdued. As the Soviet historian M. N. Pokrovsky suggested in his *Brief History of Russia* (1931):

It is easy to foresee that, in the future, when science and technology have achieved a perfection which we are as yet unable to

> visualize, nature will become nothing more than wax in our hands,
> which we are free to cast in whatever form we please.

Nature was to be reshaped to meet the needs of the Soviet people and their dominant ideology. The impact of this disastrous shift in mind-set was heightened by the displacement of older, more ecologically fecund models of nature—to which we may now turn.

The Revival of Nonmechanical Models of Nature

When I began my studies of the natural sciences, the favored method of undertaking scientific calculations was the slide rule. Science fiction television programs of the late 1960s and early 1970s showed earnest scientists in their white laboratory coats carrying out sophisticated calculations concerning the trajectories of incoming missiles. The slide rule was the height of sophistication. That was then. Nowadays the slide rule is little more than a quaint reminder of a distant past, in which things were done very differently. The electronic calculator and personal computer have made slide rules redundant. Who wants to use a slide rule when you can use a calculator? The new ways of doing things did more than make older ways obsolete; it made them seem ridiculous.

The new mechanical philosophy of nature displaced and discredited the older models of nature, above all the organic models of nature of the Renaissance. Newton's arguments in favor of universal gravitation encouraged the view that the universe was a single uniform mechanism, governed at all times and in all places by the same fundamental laws of motion. This approach to nature was profoundly hostile to the empathetic and more holistic conception of nature found in the Middle Ages and reflected in the literature of the Renaissance. As E. A. Burtt points out in his definitive *The Metaphysical Foundations of Modern Physical Science*, the new understanding of nature as a mechanism swept away earlier attempts to conceive nature as endowed with human qualities—such as "wisdom" or "harmony":

The gloriously romantic universe of Dante and Milton, that set no bounds to the imagination of man as it played over space and time, had now been swept away. Space was identified with geometry, time with the continuity of number. The world that people had thought themselves living in a world rich with colour and sound, redolent with fragrance, filled with gladness, love and beauty, speaking everywhere of purposive harmony and creative ideals—was crowded now into minute corners in the brains of scattered organic beings. The really important world outside was a world hard, cold, colourless, silent, and dead; a world of quantity, a world of mathematically computable motions in mechanical regularity. The world of qualities as immediately perceived by man became just a curious and quite minor effect of that infinite machine beyond.

It is now all too obvious that this was a premature move. What some writers of the seventeenth century regarded as a mere correction of an excessive medieval awe and respect for nature rapidly degenerated into an all-out assault on any notion of nature as a living system. What began as a *correction* to an existing worldview rapidly became an exclusivist worldview in its own right, whose rivals were dismissed as primitive superstitions which urgently required displacement by mechanical ways of thinking. Yet in reality, these organic models of nature were merely suppressed, not defeated, by the mechanical model of nature. Although the "mechanistic" analysis of reality has dominated the Western world since the seventeenth century, the "organismic" perspective has stubbornly remained in movements such as Romanticism, American transcendentalism, the German Nature philosophers, and the early philosophy of Karl Marx. The basic ideas of an organic view of nature reappeared in the twentieth century in the process philosophy of Alfred North Whitehead, to which we shall return.

The most significant early critique of the mechanical model of nature is to be found in German Romantic philosophy of the late eighteenth century, especially in the works of the tradition of plant geography inspired by

J. G. Herder (1744–1803), and Johann Wolfgang von Goethe (1749–1832) and placed on a more rigorous foundation by Alexander von Humboldt (1769–1859). We see here the development of an antimechanistic naturalism which saw nature as a dynamic process of becoming, within which humans are creative participants, not masters who simply tinker with machines. According to von Humboldt, "all the organisms and forces of nature may be seen as one living, active whole, animated by one sole impulse." Friedrich von Schelling developed this idea further, insisting that nature "is not an inert mass, and to those who can grasp her vast complexity, she reveals herself as the creative force of the universe—before all time, eternal, ever active, she calls to life all things, whether perishable or imperishable." It will be clear that these thinkers, who were deeply influenced by Romanticism, saw a fundamental causal unity at work within nature. Organisms were seen to be interdependent with each other and their environments.

The surge of interest in the "Gaia" model of nature in the late twentieth century is a telling indication of the mounting dissatisfaction with reductionist mechanical and materialist models of nature. The Gaia model has important affinities with some Christian ways of thinking about nature—again, perfectly valid ways of thinking which were improperly and prematurely suppressed by the mechanical philosophers of the Enlightenment. Yet it is important to note that James Lovelock, who was largely responsible for introducing this way of thinking, explicitly saw "Gaia" as a *model* of nature. The Gaia theory suggests that, in some sense, the earth is "alive," and proposes an organic model of nature which does justice to this aspect of its character. As Lovelock put this point:

> I recognise that to view the Earth as if it were alive is just a convenient, but different, way of organising the facts of the Earth. I am of course prejudiced in favour of Gaia and have filled my life for the past twenty-five years with the thought that Earth may be alive: not as the ancients saw her—a sentient Goddess with a purpose and foresight—but alive like a tree. A tree that quietly exists, never moving except to sway in the wind, yet endlessly conversing

with the sunlight and the soil. Using sunlight and water and nutri-
ent minerals to grow and change. But all done so imperceptibly,
that to me the old oak tree on the green is the same as it was when
I was a child.

The mechanical model stressed the reliability, orderedness, and predictability of nature, yet lost sight of any notion of nature as a living entity. Lovelock's model restores this critical dimension of our understanding of
nature. It is, of course, an understanding of nature that has always found
favor within religious traditions—for example, consider medieval Judaism's
regular use of anthropomorphic language concerning nature as "wise," and
so forth. Lovelock offers an immensely important stimulus to recover this
sidelined and suppressed way of viewing nature from its rationalist de-
bunkers.

Where older models of nature often took their inspiration from the
world of nature itself, the mechanical approach derived both its origins and
its plausibility in the world created by humanity—in the realm of human
construction rather than nature. Carolyn Merchant has identified some of
the implications and consequences of this important development:

> The rise of mechanism laid the foundation for a new synthesis of
> the cosmos, society and the human being, construed as ordered
> systems of mechanical parts subject to governance by law and to
> predictability through deductive reasoning. A new concept of the
> self as a rational master of the passions housed in a machinelike
> body began to replace the concept of the self as an integral part of
> a close-knit harmony of organic parts related to the cosmos and
> society. Mechanism rendered nature effectively dead, inert and ma-
> nipulable from without. As a system of thought, it rapidly gained
> in plausibility during the second half of the seventeenth century.

Applying N. R. Hanson's approach to these two models, we could say that
Newtonianism encouraged people to see nature as a dead, inert machine *ex-
cluding* humanity, which we could use more or less as we pleased; Lovelock's

model enables us to see nature as a living, almost sentient creature, *embracing* humanity, and thus encouraging us to treat it with respect and tenderness.

A further aspect of the rise of the mechanical worldview must be considered. The new fascination with machinery led René Descartes (1596–1650), Pierre Gassendi (1592–1655), and Marin Mersenne (1588–1648) to develop a mechanical philosophy that treated even living bodies as little more than machines.

The Ghost in the Machine: The Rise of Cartesianism

In 1690, a certain Madame de Grignon wrote to a colleague who had expressed an interest in giving her daughter a dog as a present. "Please do not!" replied this most worthy French matron. "We don't want to be burdened by a machine." Sadly, we have no information as to what happened next. However, Madame de Grignon here reflects with accuracy and enthusiasm one of the leading ideas of the Cartesian mechanical philosophy. A radical distinction had to be drawn between mind and matter. Nature was the realm of matter, which could be described in the essentially mechanical terms of extension, figure, magnitude, and motion. As Descartes stated this with his admirable clarity: "There exists nothing in the whole of nature which cannot be explained in terms of purely physical causes, totally devoid of mind and thought."

Yet Descartes held back at one critical point. Nature was a machine; dogs were machines; the human body was a machine. The human mind and soul, however, were to be treated as different. Humanity alone was to be thought of as a machine inhabited by a spiritual principle—the "ghost in the machine." At this point, Descartes faced a difficulty. If mind and matter were so radically different, how could they interact? Descartes's dualism seemed in trouble. But a creative answer was soon found. Descartes argued that the human pineal gland was the point of interaction between mind and body, allowing the human machine to be directed by its spiritual mentor. Unfortunately, it was soon discovered that dogs also had pineal glands. Did this mean that they, too, had souls and minds?

Our concern here is not to rescue Descartes from this philosophical muddle, but to note the implications of his dualism for the shaping of hu-

man attitudes to the environment. It can easily be argued that modern scientific cosmology since Newton and Descartes has helped to exile us from
nature. In that this cosmology is metaphysically based on the Cartesian dualism that exorcised mind from nature, we are left to draw the conclusion
that nature is mindless, lifeless, and purposeless material stuff. Nature is
perceived as fundamentally alien to mind, thus making it difficult for humans to feel at home in such a spiritless world. We don't belong in this
world—so why respect it?

It is hardly surprising that many reacted against this bleak view of the
universe as a vast machine. Machines could be understood—but what were
they there for? As Steven Weinberg points out in his intriguing yet problem-
atic work *The First Three Minutes* (1977), "the more the universe seems com-
prehensible, the more it also seems pointless." What for Newton was a
well-oiled clock, constructed by a wise and benevolent creator, ended up
becoming a dead and pointless mechanism, with no apparent purpose.

David Hume (1711–76), widely regarded as one of the intellectual advocates of the idea of a purposeless universe, had little doubt about its lack
of popular appeal. As C. S. Lewis pointed out in his essay "The Empty
Universe," the notion was widely regarded as unbearable, even by Hume
himself.

> [Hume] recommended backgammon instead, and freely admitted
> that, when, after a suitable dose, we returned to our theory, we
> should find it "cold and strained and ridiculous." And obviously, if
> we really must accept nihilism, that is how we shall have to live,
> just as, if we have diabetes, we must take insulin. But one would
> rather not have diabetes, and do without the insulin. If there
> should, after all, turn out to be any alternative to a philosophy that
> can be supported only by repeated (and presumably increasing)
> doses of backgammon, I suppose that most people would be glad
> to hear of it.

It is hardly surprising that culturally enlightened and spiritually sensitive individuals of the late eighteenth and early twentieth centuries reacted against

Hume's ideas, as we shall see when we consider the demand for the reenchantment of nature from within the Romantic movement.

Our attention now turns to an antimechanical philosophy developed
by a leading twentieth-century Jewish philosopher, which represents an important reaction against mechanical approaches to nature.

Nature as "Thou": Martin Buber

How does one relate to God, to other people, and to nature? The noted
Jewish philosopher Martin Buber (1878–1965) sought to provide a work-
able answer to these questions in his famous essay *I and Thou* (1927). This
work draws upon some basic themes of Jewish thought, especially in relation to how God is experienced and known, which many Christian writers
have subsequently found helpful and constructive. Buber had long been dissatisfied with the weakness of the Enlightenment approach to nature, which
invariably portrayed nature as an "It," an inanimate object investigated by
active human subjects. In this seminal work, Buber provides a typology to
describe two types of relationships into which human beings enter. According to Buber, human beings possess a twofold attitude toward the world in-
dicated by the primary words "I-It" and "I-Thou."

In the "I-It" relationship, we *experience* something. This distinction is
often referred to in more philosophical language as a *subject-object* relation, in
which an active subject (such as a human being) relates to an inactive object
(such as a pencil). According to Buber, the subject here acts as an "I" and
the object as an "It." The relation between the human being and pencil
could thus be described as an "I-It" relation. This type of relationship is
characterized by the objectification and control of nature and people. The
"I" in this relationship seeks to acquire and possess as much as it can and
perceives itself as being an individual, who is set over against the subjects of
its perception. But this "I" pays a price for such selfishness and will to
dominate because it is isolated and alienated from the source of life. What
also characterizes the "I-It" relationship is that it is embedded in space and
time and determined by causality. This relationship includes mundane acts
such as mass consumption, industrial production, and societal organization.

It seems clear from Buber's text that the world of technical mechanization and scientific objectification and control results from the "I-It" model of apprehending reality. Buber's aversion to modern technology arises from his fundamental conviction that it has contributed to the expansion of the domain of "I-It" and the diminishing of the "I-Thou" domain. The modern self is more likely to relate to the "other" as an "It" than as a "Thou." This is what Buber means when he speaks of humanity descending into alienation in the modern industrial world. Technology fosters the perception that the "other" is an "It," which can be shaped as humanity pleased and with which a relationship is inconceivable. To treat nature as "It" is to engender an ethic of alienation, exploitation, and disengagement.

In marked contrast, an "I-You" relation designates the world of *encounter*. Such a relation exists between two active subjects—between two persons. It is something that is *mutual* and *reciprocal*. "The I of the primary word I-You makes its appearance as a person, and becomes conscious of itself." In other words, Buber is suggesting that human personal relationships exemplify the essential features of an "I-You" relation. It is the relationship itself, that intangible and invisible bond that links two persons, that is the heart of Buber's idea of an "I-You" relation. Buber sees this as a more authentic and deeper way of relating to the world. To know nature fully, we must see it as a "Thou" rather than an "It." We must *encounter* nature, not simply *experience* it.

According to Buber, "I-It" knowledge is indirect, mediated through an object, and has a specific content. In contrast, "I-Thou" knowledge is direct, immediate, and lacks a specific content. An "It" is known by measurable parameters—its height, weight, color, and so on. But a "You" is known directly. To treat nature as an "It" is to see nature merely as something that can be analyzed and exploited by humanity rather than something with its own integrity and existence.

Buber does not argue that nature is *intrinsically* possessed of qualities that demand that it be treated as a "Thou." Rather, he shows how humanity can relate to nature in a respectful manner on account of its *perception* of the intrinsic value of nature, whether such intrinsic value actually exists. Yet it is clear that, if it could be shown or argued that nature did indeed possess such qualities, the case for treating it as a "Thou" would be considerably strengthened. It is at this point that Buber's Jewish roots provide him with the insight needed to clinch the argument—namely, that nature does indeed possess such mystical qualities, on account of its relationship with God.

Where do we go from here? One vitally important point that must be made immediately is that we need to affirm the importance of deploying multiple models of nature. The developments documented in this chapter rest on the uncritical acceptance of one model as if it were complete in itself or as if there were no other viable models available. This is most emphatically not the case. The mechanical model allows us to visualize nature in a way that helps us appreciate its ordering, regularity, and predictability. Yet it fails to do justice to other aspects of the natural order, such as those noted by Lovelock. The mechanical model illuminates *part* of the complex entity we know as nature; the problems arise when we think it describes it *totally*.

There is no doubt that the rise of the mechanical view of the world disenchants nature and deprives it of any special relationship to humanity. In her *Death of Nature*, Carolyn Merchant argues that before the scientific revolution, nature was seen as a "nurturing mother" or as "a kindly beneficent female who provided for the needs of mankind in an ordered, planned universe." This powerful and dominant metaphor of the earth as a nurturing mother was cast aside in favor of the mechanical and rationalist worldview of the Enlightenment, which in turn gave rise to a "mechanical model of society as a solution to social disorder." Within this conceptual framework, natural scientists were seen as those who analyzed the world, manipulated it, and provided the justification to manage, control, and exploit nature. The outcome of this, Merchant argued, was inevitable: no one considered the permanent insult to the earth caused by such things as strip mining, dumping waste into rivers, clearing vast tracts of forests, and draining marshes.

Merchant explains how a growing ability to alter the face of the earth had a significant impact on the way in which people viewed their relationship with the earth. Before the scientific revolution, the "earth was considered to be alive and sensitive." As Western culture became increasingly mechanized in the 1600s, the earth was subdued and depersonalized by machines. The growing power of machines eventually led to nature itself being seen in terms of its conquerors—as a mechanism.

There is no need to revert to a premodern worldview to redress this serious imbalance in the way we now see nature. To see nature as a mechanism is to recognize and celebrate its regularities and predictabilities, upon which we ultimately depend for our existence, as ancient Egypt depended on the regular flooding of the Nile for its food. Yet this illuminates only one element of nature and cannot be allowed to determine its every aspect. We need to recover alternative models, illuminating additional aspects of nature, many of which have their roots deep in the Christian tradition, and rediscover their relevance to our ecological crises. This process of retrieval does not negate the scientific investigation of nature, or place any barriers in the way of the continuing attempt to understand its complexities. It simply argues that there is more to nature than a mechanism placed there for our convenience.

To develop this point, we may move on to consider the very different understandings of nature associated with the Enlightenment and with Romanticism.

CHAPTER SIX

Dissatisfaction with Spiritual Emptiness: The Longing for Reenchantment

In my view, the natural sciences offer the most satisfying intellectual engagement that humanity can ever hope to experience, surpassing anything I have ever known throughout my subsequent career as a researcher in history, literature, and religion. The uncovering and explanation of the structured depths of nature continue to fascinate and excite me, evoking a sense of delight and wonder shared by countless others working in these fields. Yet I write these words as a wounded lover, one who has become increasingly aware of a darker side of what I had once seen in a solely positive light.

Today the proclamation of the total competence of the natural sciences finds a muted welcome within Western culture and invites a more systematic rebuttal. Whether for better or worse, fairly or unfairly, science is increasingly coming to be seen as a negative influence upon human culture. Where our grandparents saw science as offering solutions to our problems, many of our generation are seeing it as the source of new problems. The British scientist Sir Richard Gregory (1864–1952) proposed the following as his epitaph:

My grandfather preached the gospel of Christ
My father preached the gospel of socialism
I preach the gospel of science.

Gregory was not out to rubbish religion or politics, both of which he regarded as far too important to be treated in such a dismissive manner. Science, he insisted, was to be regarded "as one of the great human endeavors to be ranked with arts and religion as the guide and expression of man's fearless quest for truth." He simply saw the sciences as having displaced older sources of hope for humanity. Where others looked to religion or politics for security and hope for the future, he believed that future generations would look to the sciences. Today that judgment would be subjected to considerable criticism.

Two Views of Nature: The Enlightenment and Romanticism

A new generation has arisen that—rightly or wrongly—attaches significant blame to the natural sciences for the ills of our age. The sciences are critiqued for facilitating the exploitation of the earth and the pillaging of its resources. They are seen as having encouraged the rise of materialism and the depreciation of spirituality. This is perhaps unfair, as it fails to make the critical distinction between those who study nature out of sheer fascination and those who study it in order to tame and master it. One of the more disturbing aspects of human nature is that the knowledge that some gained from sheer intellectual delight can be taken and used for less reputable purposes by others. It is impossible to control the use to which knowledge is put.

In their highly significant book *Contested Natures* (1998), Phil Macnaghten and John Urry explore the attitudes toward nature that have emerged since the seventeenth century. As a result of their close analysis, they discern two broad patterns of approaches which, they argue, cast their long shadows over contemporary attempts to grapple with the meaning and significance of the natural world. The two cultural frameworks they identify are termed the *Enlightenment* view, which adopts a "nature-skeptical" attitude, and the *Romantic* view, which is distinguished by its "nature-affirming" outlook. Macnaghten and Urry point out that these two very different ways of experiencing, perceiving, representing, and valuing nature are laden with significance in terms of the quality and significance of the human encounter with the natural order.

For Macnaghten and Urry, the Enlightenment saw nature in terms of a struggle for power and survival, "red in tooth and claw" (Tennyson), which regarded life as "solitary, nasty, brutish and short" (Hobbes). Nature is thus a primeval state that must be mastered and transformed in order to be converted into something more civilized and useful to humanity. This necessitated the development of technology through the natural sciences, which allowed the crudities of nature to be transmuted into commodities and possibilities more acceptable to humanity.

Although Macnaghten and Urry overlook the matter, there is a fascinating point to be made in relation to the changing meaning of the word "urbanization" in the early nineteenth century. The word initially had the sense of the "process of making urbane"—that is, a process of social refinement, leading to the acquisition of certain genteel qualities much prized by English cultured *bien pensants* of the time. By the middle of the century, the word had shifted its meaning and begun to acquire more questionable associations. It is at this point that the meaning of the "conversion of rural to urban," with its subtext of the mechanistic agenda of the Industrial Revolution, makes its appearance. The Enlightenment, it can easily be argued, began with the entirely praiseworthy intention of civilizing the crudities of nature; it ended up destroying England's pastoral economy, turning the countryside into a vast disease- and poverty-ridden urban sprawl.

In contrast, the Romantic view took as its starting point the belief that humanity and nature were once in harmony and that this innocent relationship was disrupted and ultimately all but destroyed through scientific progress and the mechanization of the world. What the Enlightenment termed "civilization" is actually a process of alienation of humanity from its proper habitat and the physical destruction of the environment in which humanity ought to exist.

In view of the importance of Romanticism in this respect, we shall consider it in a little more detail.

The Dream of Romanticism: Glimpsing the Transcendent

To appreciate the distinctive character of Romanticism, it is necessary to set the context in a little detail. The eighteenth-century Enlightenment generated a heritage of optimism about human possibilities and above all human perfectability. The Enlightenment portrayed nature as something that can be understood by human reason as a precondition for mastering and recasting it by human hands. We have already seen the devastating impact of this worldview under Stalin, and its impact on the ecology of the Soviet Union during the 1930s. Aware that religion offered a significant barrier to this process of exploitation, the Enlightenment offered a vigorous critique of traditional religious ideas, especially those of Christianity, within whose heartlands the movement emerged. While this criticism weakened the intellectual appeal of Christianity, the alternative that the Enlightenment proposed was less than appealing to many.

The aspect of the Enlightenment project of particular importance to us is the mechanical view of the world it proposed. As we have seen, the Newtonian and Cartesian mechanical universe left many people deeply dissatisfied, not simply those with religious views. The mechanistic view of the universe was increasingly rejected, not so much on rational as on emotional and intuitive grounds. There had to be something better, something more fulfilling. Many opted for a more organic view of nature, seeing the world as a dynamic living organism rather than an impersonal mechanism. The new material dogmas of reason seemed dry and cold and were unable to satisfy the deep spiritual yearnings that continued to burn within many. While not necessarily wishing to endorse the traditional ideas of Christianity, many were looking for spiritual roots that would give their lives dignity, meaning, and purpose. They turned to nature.

Nature was now seen in vastly elevated terms as the moral and spiritual educator of humanity. This romanticization of nature began in earnest in the late eighteenth century, initially in the writings of German Romantics such as Goethe and Novalis. The trend is probably seen at its most pronounced in English Romanticism. A good example is found in the famous lines from William Wordsworth's "The Tables Turned" (1798):

> One impulse from a vernal wood
> May teach you more of man,
> Of moral evil and of good,
> Than all the sages can.

In his later works, Wordsworth develops the theme of the ability of the natural world to evoke an aching sense of longing for something that ultimately lies beyond it—as in "Tintern Abbey," which uses the poet's experience of a natural landscape to evoke deeper questions about the mystery of human nature and destiny.

> The sounding cataract
> Haunted me like a passion: the tall rock,
> The mountain, and the deep and gloomy wood,
> Their colours and their forms, were then to me
> An appetite: a feeling and a love.

There is a strong sense of the loss of connectedness here, a deep and passionate feeling that individuals have become alienated from nature.

A similar theme can be found in other writings of the period fostering sentimentalism about nature, which often suggest that arcadian nature—as opposed to the pillaged and mutilated version we now know—is the only responsible moral education for modern youth. Where science and capitalism mislead and destroy, nature leads its followers onward to personal fulfillment and integrity. Percy Bysshe Shelley expressed this idea in his "Hymn to Intellectual Beauty," which posits the idea of an intuited higher power, which saturates nature with its presence and beauty:

The awful shadow of some unseen Power
Floats though unseen among us,—visiting
This various world with as inconstant wing
As summer winds that creep from flower to flower.

The human experience of this beauty may be sporadic rather than continual; it is nevertheless an integral aspect of the phenomenon of nature. Nature is not simply to be investigated and understood as the "other" by detached observers; it is to be encountered and it is to evoke wonder at its sheer beauty by a humanity that is aware that it is an active participant of the whole great scheme of things rather than a detached and uninvolved observer.

The Romantic view of nature makes points of fundamental importance, which need to be taken with great seriousness by anyone concerned with the proper human evaluation and perception of nature. Having said that, there are some obvious problems with the ideas we find in, for example, Wordsworth. The most obvious is a failure to appreciate the extent to which the "natural" landscapes and vistas that Wordsworth so prized are actually the outcome of the human transformation of nature. Wordsworth disliked the intrusion of debasing human activities into otherwise pastoral scenes. Many have praised the natural landscapes of France, Greece, and New England, seeing in them some kind of pristine purity. Others have sought inspiration in the great natural wonders of North America, which display nature at its finest—such as Yellowstone National Park, the Mississippi River, and Niagara Falls. Such landscapes and natural wonders are portrayed as independent of humanity, possessing an integrity and simplicity that contrasts with the crude artifacts of humanity. Yet the reality of the situation is actually rather different. These "natural" landscapes and features are socially mediated. The English natural landscapes that evoked so powerful a reaction in Romantic poets such as Wordsworth and Coleridge had been tamed by centuries of human presence and agricultural work, and derive at least part of their charm from picturesque buildings—primarily churches—which, of course, are human constructions. The so-called natural landscapes of New England and Europe are the result of human habitation and transformation, with the imprint of human civilization evident at point after point.

Similarly, the features of the landscape of the great American West owes far more to human activity than is acknowledged by those who long for nature at its purest; the nature whose beauty is admired by the Romantics can be argued to be a nature that has been tamed and transformed—although in culturally pleasing ways—by human agency and activity.

Despite such matters, many continue to find nature as a category that provokes a discourse of transcendence. There seems to be something about the vastness of the natural world that forces questions upon us, often of a metaphysical nature. In what follows, we shall begin to interact with some of these.

The Transcendence of Nature

The movement that is still known as the American Renaissance is of especial importance to our themes. Although many of the themes of this movement drew their inspiration from its European counterpart, they were shaped and transformed by the distinctives of the American situation—above all, by a sense of the vastness of nature. This strongly idealized conception of nature played a major role in the shaping of American attitudes to the wilderness from which an emerging nation was being hewed. As Perry Miller points out in a classic study of the shaping of American concepts of and attitudes toward nature, the concept of America as "nature's nation" emerged in the nineteenth century, along with a belief that the natural order could educate and safeguard the moral standards of the new nation. Writing in January 1840, Walt Whitman saw nature as providing the inspiration for his nation, then struggling to find and retain its distinctive identity on the global stage:

America can progress indefinitely into an expanding future without acquiring sinful delusions of grandeur simply because it is nestled in Nature, is instructed and guided by mountains, is chastened

by cataracts . . . So then—because America, beyond all nations, is
in perpetual touch with Nature, it need not fear the debauchery of
the artificial, the urban, the civilized.

Yet the growing American appreciation of nature was linked with the
perception that it was being lost just about as fast as people were coming to
appreciate it. The "pioneer spirit" meant that an increasing number of
physical frontiers were being conquered in this time of "manifest destiny";
as a result, there was less and less wilderness to explore. Many American
Romantic writers turned their attention to artistic, metaphysical, and intellectual frontiers, as they attempted to recapture the sense of ecstasy of exploration and discovery, without causing any damage to the environment.

This strong sense of awe in the presence of nature, found in
Wordsworth's best nature poems, is characteristic of many of the writings
of Henry Thoreau (1817–62). His period spent at Walden Pond
(1845–47) led him to gain a deep awareness of the connection between
humanity and nature. This period of "deliberate living" also led him to
write *Walden,* one of the best-known works of nineteenth-century American
literature. His delicious prose and his descriptions of his interactions with
nature testify to his firm belief that there is an unseen power just beyond
the veil of the visible. For Thoreau, humanity stands in the midst of some
deep mystery which unadulterated nature lifts aside from time to time, allowing us access to its secrets. It is only through developing a proper relationship with nature that individuals and communities can achieve true
fulfillment.

Perhaps the most vivid evocation of the sacred intellectual and spiritual
depths of nature to emerge from the American Renaissance is to be found
in Ralph Waldo Emerson's essay "Nature" (1836). In this essay, he tried to
express in words a profound awareness of the transcendence of nature. For
Emerson, the outward world was only an appearance or dream, and had no
real substance. It was the manifestation of the spiritual world—the solidi-
fied or embodied thoughts of God. Why, asked Emerson, did we need to
read the writings of long-dead theologians to gain access to the mind of
God when it could be seen around us, in the sacred depths of nature?

The foregoing generations beheld God face to face; we, through their eyes. Why should not we also enjoy an original relation to the universe? Why should not we have ... a religion by revelation to us, and not the history of theirs? Embosomed for a season in nature, whose floods of life stream around and through us, and invite us by the powers they supply, to action proportioned to nature, why should we grope among the dry bones of the past?

This point can be illustrated by the use of the image of a river. Practical Americans of the mid-nineteenth century, such as those Emerson knew through his addresses at the Boston Mechanics Institute, viewed rivers primarily in utilitarian terms. Rivers posed mechanical challenges (such as how to build bridges over them or control their flux) and opportunities (such as providing drinking water, flushing wastes, and affording transport). Emerson, however, was not particularly interested in these physical uses of rivers. For him, the river was an image of beauty and a source of spiritual energy, connecting the universe into a fluid whole.

"Every natural fact is a symbol of some spiritual fact." The human imagination, according to Emerson, is raised to new heights through the stimuli it receives from nature. Nature acts as a signpost to the transcendent, evoking an awareness of the sublime hand of God which fashioned the creation:

> One might think the atmosphere was made transparent with this design, to give man, in the heavenly bodies, the perpetual presence of the sublime ... If the stars should appear one night in a thousand years, how would men believe and adore; and preserve for many generations the remembrance of the City of God which had been shown! ... But all natural objects make a kindred impression, when the mind is open to their influence.

It is, of course, easy to dismiss Emerson as something of a crank, a nutcase who projects onto nature the secret longing for meaning that he nurtures within his breast. There is no meaning or purpose in nature, it might be argued, save what Emerson ludicrously invents. What Emerson claims to discern within nature has in fact been imposed upon it—and imposed in an arbitrary and illogical manner. Yet Emerson stands as a powerful witness to the human refusal to accept that the Newtonian world is all that there is. *There has to be more*. The human relationship with nature is not that of detached observer, but that of an active participant. Emerson makes this point as follows:

> To the senses and the unrenewed understanding, belongs a sort of instinctive belief in the absolute existence of nature. In their view, man and nature are indissolubly joined. Things are ultimates, and they never look beyond their sphere. The presence of Reason mars this faith. The first effort of thought tends to relax this despotism of the senses, which binds us to nature as if we were a part of it, and shows us nature aloof, and, as it were, afloat. Until this higher agency intervened, the animal eye sees, with wonderful accuracy, sharp outlines and colored surfaces. When the eye of Reason opens, to outline and surface are at once added, grace and expression. These proceed from imagination and affection, and abate somewhat of the angular distinctness of objects. If the Reason be stimulated to more earnest vision, outlines and surfaces become transparent, and are no longer seen; causes and spirits are seen through them. The best moments of life are these delicious awakenings of the higher powers, and the reverential withdrawing of nature before its God.

This approach to nature reached its zenith in the poetry of John Keats (1795–1821). In his 1820 poem "Lamia," Keats complains of the effect of reducing the beautiful and awesome phenomena of nature to the basics of scientific theory. The theory may help us understand them—but somehow it seems to deprive them of their glory. Keats here expressed a widely held concern: that reducing nature to scientific theories empties nature of its beauty and mystery and reduces it to something cold, clinical, and abstract.

Do not all charms fly
At the mere touch of cold philosophy?
There was an awful rainbow once in heaven:
We know her woof, her texture; she is given
In the dull catalogue of common things.
Philosophy will clip an Angel's wings.

Keats uses the idea of "unweaving the rainbow" to express his concern. Does not the scientific explanation of the colors of the rainbow in terms of refraction of light through raindrops destroy any sense of awe or amazement at this arc in the sky? Nature has become reduced to quantities.

There is an important point here: Keats uses the term "philosophy" as a kind of shorthand for "natural philosophy," which is how the English academic establishment preferred to speak of what we would now call the natural sciences. The habit continued. When I was studying chemistry at Oxford in the early 1970s, the scientific disciplines were still referred to as natural philosophy. Oxford is terribly good at preserving past traditions, even when everyone has forgotten what they once meant.

Keats's protest thus goes far deeper than is usually appreciated, and makes a point of fundamental importance. *Nature offers us glimpses of the transcendent.* It is a shadow of a greater and brighter reality, which will one day be attained or encountered. To understand this point, and why it matters so profoundly to our theme, we shall move on to consider it in more detail, before turning to a recent attempt to refute Keats by the British biologist Richard Dawkins.

The Wonder of Nature and Intimations of Glory

The Christian tradition has seen the created order as pointing to its creator. This is particularly the case within Greek Orthodox theology, which sees the entire universe as an eloquent expression of the Word of God—a "cosmic liturgy" as Maximus the Confessor described the world in the seventh century. Christian theology argues that the sense of wonder that we experience on beholding the night sky, a breathtaking sunset, a glorious mountain vista, or—to use perhaps the most overworked example of all—a rainbow has three aspects, each of which contributes to our appreciation of the beauty of nature.

First, there is an immediate sense of awe, which is not mediated through any understanding or cognitive agency. Wordsworth expressed this immediate awareness of beauty like this:

My heart leaps up when I behold
A rainbow in the sky.

Even before the mind grasps the scientific principles that govern the shape and colors of that rainbow, the human heart has responded at a far deeper intuitive level. As Blaise Pascal once noted, "the heart has its reasons, which the mind does not understand." For the Christian theologian, this powerful and immediate response of the human heart to beauty is grounded in the creation of both humanity and nature by God, which points to a fundamental resonance between human perception and natural beauty.

Second, there is an enhanced sense of appreciation that results from understanding how the beauty of nature comes about—for example, Newton's optical explanation of the colors of the rainbow. Understanding need not destroy the beauty of nature. But—and here Dawkins needs to be challenged—it *can*. Dawkins represents the outlook of the Enlightenment, which placed an immense weight on the rational activities of the human mind and deemphasized the role of the imagination and emotions. As countless critics of the Enlightenment have pointed out, it is not sufficient to appeal to the human mind; it is essential to capture the imaginative faculties of that human mind. Newton's theory of optics is, of course, interesting, and countless generations of children have been duly inculcated with its principles. But once you've grasped the principles, it can easily lose its fascination and become a little dull. As Samuel Johnson (1709–84) once quipped of well-intentioned scientific educationalists: "You teach your daughters the diameters of the planets, and wonder when you are done that they do not delight in your company."

Third, the wonders of nature point to the glory of God. Nature is able to direct our attention to something more glorious that lies beyond it. This idea is fundamental to Christian theology and can be discerned in modified forms at countless junctures in literature. For example, the New England Transcendentalists were firmly committed to the idea that the beauty of nature disclosed something of the realm of the transcendent, which otherwise might be inaccessible to us. Wordsworth's nature poetry is often described in terms of its "liminality"—a sense of standing on the border of some unknown territory, which is signposted by what we know, yet which ultimately lies beyond it.

According to this view, nature is not complete in itself. It points to something, intimating the presence or promise of something even more wonderful. To study and understand nature is an excellent thing; to appreciate that something else lies beyond it does not in any way diminish the quality of the human engagement with nature in poetry, art, or science; if anything, it heightens the motivation for that quest. The greater our appreciation of the glory of nature, the greater our longing for a transcendent glory that lies beyond it but that is faintly reflected in our own world of experience.

Some aspects of these issues have been explored in Philip Fisher's *Wonder, the Rainbow, and the Aesthetics of Rare Experiences* (1998). In this work, Fisher argues that the sciences have not reduced the sense of wonder that many experience in encountering nature—for example, a rainbow. The rainbow, he points out, has always provoked people to offer an explanation of its beauty. Aristotle offered one explanation in the classical period, Theodoric of Freiburg in the Middle Ages, and Isaac Newton in more recent times. Newton was even able to intensify the wonder of the rainbow by explaining the phenomenon of the "double rainbow." Fisher thus concludes that, far from destroying the wonder of the rainbow, science actually enhances it. He contrasts this with religious approaches to nature, which he argues impoverish that sense of wonder. Where science expands the wonder of nature, religion forecloses any future exploration of the matter.

I have no quarrel with Fisher's suggestion that science enhances the wonder of nature, and suspect that many will find this unproblematic. The more we appreciate the intricacy and complexity of nature, the greater our sense of awe at its grandeur, and satisfaction at being able to grasp at least something of its structure. Yet it is necessary to protest against his highly implausible (mis)reading of religious approaches to nature, which seem to miss the point entirely (although in a rather elegant and sophisticated manner).

A religious view of nature denies nothing of this sense of wonder, nor does it seek to limit its exploration in any way. Aristotle once commented that "philosophy begins from a sense of wonder," and Christian theology would readily endorse such a suggestion. Yet that sense of wonder is extended through the fundamental theological affirmation that nature is a sign—that it points to something even greater and that at least something of this greater wonder may be known through the natural world. The Christian reading of nature issues an "invitation to ascend"—to borrow a phrase from Stanley Fish's excellent study of Milton's *Paradise Lost*, which Fish interprets as describing a "Platonic ascent" which leads to "the simultaneous apprehension of the absolute form of the Good and Beautiful." Nature points upward, aesthetically and ontologically, and we fail to appreciate its richness until we become caught up in this upward trajectory. To observe the stars of the heavens is one thing; to see them as signs of a beauty and goodness that is yet to be fully revealed—but that we may *anticipate* now—is quite another.

In 1975, I traveled to Iran with a colleague, leaving behind a damp British summer. We spent several weeks touring the country. The heat of the sun was such that it was impossible to travel between major cities by bus during daylight hours. The Iranian bus companies arranged their schedules so that long-distance travel took place in the cool of the night. Buses would leave cities such as Isfahan for Shiraz at about ten at night and arrive at their destinations around dawn, before the day got unbearably hot.

We set out from the fabled city of Shiraz one night for the eastern city of Kerman, close to the border with Pakistan. The skies were totally cloudless, and as the night progressed, the stars seemed to shine with an almost supernatural brilliance. As the bus neared a deserted caravanserai, it became clear that there was a problem. Its ancient diesel engine began to make ominous coughing noises. The bus finally ground to a halt, spluttering alarmingly while the driver tried to work out what was the matter.

The forty or so passengers got off and wandered around the ruins of the caravanserai. It had once been an important staging post for traders in the nineteenth century, and some were curious to explore the site while we waited for the driver to fix the engine. I found myself transfixed by something rather different. Having lived for many years in English cities with high levels of light pollution, I had rarely witnessed the spectacular glory of the night sky. The lights of English cities prevent the sky from ever becoming really dark at night, and the high level of atmospheric pollution prevents the stars from being seen in their full glory. In the stillness of a remote and silent desert, however, the stars were astonishingly brilliant. I was overwhelmed with a sense of wonder. I had never seen a sight like this before. It is easy to see how this immediate sense of wonder could lead someone to want to spend his or her life studying the stars, longing to know more about their mysteries.

I had often admired the poems of the American writer Walt Whitman, especially those published in the slim volume *Leaves of Grass*. The best known of these contrasts the explanations and theories of the professional astronomer with the direct experience of the night sky:

> When I heard the learn'd astronomer;
> When the proofs, the figures, were ranged in columns before me;
> When I was shown the charts and the diagrams, to add, divide, and measure
>   them;
> When I, sitting, heard the astronomer, where he lectured with much applause in
>   the lecture-room,
> How soon, unaccountable, I became tired and sick;
> Till rising and gliding out, I wander'd off by myself,
> In the mystical moist night-air, and from time to time,
> Look'd up in perfect silence at the stars.

I could understand something of what he meant. A direct encounter with nature is always going to trump a tedious lecture about nature. Yet Whitman seemed quite wrong in one matter. Knowing something about astronomy did not diminish the glory of that night sky. If anything, it enabled me to appreciate it all the more, not least through impressing upon me a sense of the immensity of the universe.

The question of what those stars signified cannot be overlooked. Some would argue that they signify precisely nothing. They are just there, and that is all there is to be said. Anything else is pure speculation. Yet the Christian doctrine of creation affirms that something of the glory of God can be known and experienced through gazing at the wonders of the night sky: "The heavens are telling the glory of God; and the firmament proclaims his handiwork" (Psalm 19:1). My experience of the brilliance of the night sky was thus significantly extended on account of the lens through which I viewed it. In addition to appreciating the intrinsic wonders of the heavens, they were pointing me toward ideas that also evoked wonder—such as glimpsing something of the glory of God. Both the sign in itself and the greater reality to which it pointed had the capacity to evoke a sense of wonder within me. It is little surprise that so many natural scientists are religious, when they find their faith being reinforced and given greater depth through their engagement with the wonders of nature, which invites them to engage on an upward imaginative trajectory. But where does this lead us?

Heaven in Ordinary: The Concept of Transsignification

In one of the poems making up the remarkable collection known as *The Temple*, George Herbert deploys a powerful series of phrases and images to enable his readers to bring their imaginations to bear on the idea of prayer. The images he jumbles together illuminate the many aspects of prayer within the Christian tradition, not least in evoking a silent wonder in the presence of God:

> *Softness, and peace, and joy, and love, and bliss,*
> *Exalted manna, gladness of the best,*
> *Heaven in ordinary, man well drest,*

The milky way, the bird of Paradise,
Church-bells beyond the stars heard, the soul's blood,
The land of spices; something understood.

One phrase serves our purposes in this section especially well: "heaven in ordinary." In some sense, the objects and events of this present world are able to function as an intimation of eternity, almost as if they cast aside a veil allowing us a brief glimpse of God.

Within the Christian tradition, this idea is particularly associated with the bread and wine used throughout the Christian community to recall and remember the death of Christ. In some way, ordinary things—such as bread and wine—come to possess a deeper significance, pointing beyond their present status to allow us to anticipate a future revelation of the glory of God, and entering into the divine presence. The origins of this Christian practice go back to Jesus Christ himself and are based on the Gospel accounts of the Last Supper:

> While they were eating, [Jesus] took a loaf of bread, and after blessing it he broke it, gave it to them, and said, "Take; this is my body." Then he took a cup, and after giving thanks he gave it to them, and all of them drank from it. He said to them, "This is my blood of the covenant, which is poured out for many. Truly I tell you, I will never again drink of the fruit of the vine until that day when I drink it new in the kingdom of God." (Mark 14:22–25)

The basic idea is that in some way bread and wine signify or point to the body and blood of Christ. But how?

One answer to this question was given in the sixteenth century by the Swiss Protestant theologian Huldrych Zwingli. Zwingli held that the fundamental process was a change in what was being signified. The bread, which once signified physical food, comes to possess a deeper significance. Zwingli's argument runs like this: What makes the bread at a communion service different from any other bread? If it is not the body of Christ, what is it?

Zwingli answers this question with an analogy illustrating how a change in signification comes about. Imagine a queen's ring, he suggests. That ring can be considered in two quite different contexts. In the first context, the ring is physically present in a neutral context—for example, it might be lying on a table. In this situation, it has no particular associations. The ring signifies a piece of precious metal, and nothing more. Its significance is restricted to its material identity; there is no "added layer" of meaning. Now imagine precisely that same ring, but transferred to a different context. It is placed on the finger of the queen, as a gift from her king. Zwingli argues that the same ring now has personal associations, deriving from its connection with the king—such as his authority, power, and majesty. Its value is now far greater than that of the gold of which it is made. It possesses a new stratum of significance, which transcends its material identity as a piece of precious metal. These new associations arise through transfer from the original context to the new context. The ring itself remains completely unchanged.

If signification is dependent solely upon the material identity of the ring, it remains unchanged throughout. But Zwingli's point is that the signification of a physical entity—such as a golden ring or a piece of bread—is shaped by its context, or by the interpretation placed upon it by those who regard it as being significant.

This is the case with the bread and wine of a communion service, Zwingli argues. When bread or wine is moved into this new context, it takes on new and important associations. Above all, when bread and wine are placed at the center of a worshiping community, and when the story of the last night of the life of Christ is retold, they become powerful reminders of the foundational events of the Christian faith. It is their context that gives them this meaning; they remain unchanged in themselves.

Zwingli did not use the term "transsignification" to refer to his theory of the changed meanings of physical objects through the beliefs of those who behold them. The term was first used extensively during the 1960s, particularly by a group of Belgian Roman Catholic theologians who found themselves ill at ease with the traditional language of "transubstantiation." In his important study *The Eucharist* (1968), Edward Schillebeeckx argued that the interpretation of the signification of the bread and wine as the body and blood of Christ is not arbitrary, nor is it a human imposition upon them; it is an act of discernment by the church, which has been *authorized* to make this connection by Christ. In other words, this shift in significance is not imposed by human beings, but is authorized by Christ himself.

There is, for Schillebeeckx, no need to invoke the notion of a physical change of substance of the bread and wine. Christ's intention was not to alter the metaphysics of the bread and wine, but to ensure that these pointed to his continuing presence within the church, as the community of the faithful.

> Something can be essentially changed without its physical or biological make-up being changed. In relationships between persons, bread acquires a sense quite other than the sense it has for the physicist or the metaphysician, for example. Bread, while remaining physically what it was, can be taken up into an order of significance other than the purely biological. The bread then *is* other than it was, because its determinate relationship to man plays its part in determining the reality about which we are speaking.

The parallels with Zwingli's position, though not noted by Schillebeeckx himself, can hardly be overlooked.

It is thus possible to view nature *as* a pointer to the glory of God, without requiring any change in the material reality of nature involved in this radically different perception. Nature remains what it was; yet its signification—what we understand it to point to—has changed utterly. There is nothing "magical" or irrational about this. It is about a change in the way in which the world is perceived.

So how does this excursion into some of the finer points of Christian theology help us understand how nature can act as a sign? Perhaps the easiest way to grasp the point at issue is to reflect a little on my own experience as a convert to Christianity, in which I experienced a radical change in my understanding of the signification of one specific aspect of nature—the stars at night. Before discovering Christianity, I had seen the stars of the heavens as heightening our sense of transience and finitude, forcing us to ask whether this life is all that we can hope for. My growing knowledge of astronomy helped me appreciate the beauty of the universe. Yet it was a deeply melancholy beauty, in that I was unable to detach the glory of the heavens from the transience and fragility of the one observing that glory.

It was as if the stars proclaimed the insignificance and transience of those they allowed to observe them. I was totally in sympathy with the ideas I found in *The Rubáiyát of Omar Khayyám,* a classic work of Persian literature, which gives powerful expression to the deep sense of despondency evoked by the heavens. We are powerless to change our destiny. The sun, moon, and stars declare both our transience and apparent inability to change our situation.

> *And that inverted bowl we call "the Sky,"*
> *Whereunder crawling cooped we live and die,*
> *Lift not thy hands to It for help—for It*
> *Rolls impotently on as Thou or I.*

I thus saw the stars as a melancholy reminder of the vastness of the universe and the utter significance of humanity within it. However, as this was the way things were, I had no problem in accepting it. It wasn't especially attractive, but I somehow had to make the most of it.

That sort of thought has gone through the minds of many natural scientists and is particularly well expressed in Ursula Goodenough's reflective book *The Sacred Depths of Nature* (1998). As one of North America's leading cell biologists, Goodenough recalls how she used to gaze at the night sky, reflecting on what she observed. Each of the stars she saw was dying, as would our own special star, the sun. "Our sun too will die, frying the Earth to a crisp during its heat-death, spewing its bits and pieces out into the frigid nothingness of curved spacetime." She found such thoughts to be overwhelming and oppressive: "The night sky was ruined. I would never be able to look at it again . . . A bleak emptiness overtook me whenever I thought about what was really going on out in the cosmos or deep in the atom. So I did my best not to think about such things."

I felt exactly that same sense of melancholy and devised more or less the same coping plan. It was best not to think about the pointlessness of life. One of those who lectured to me on quantum theory at Oxford at this time was Peter Atkins, a physical chemist with a strong commitment to atheism. He would later write as follows concerning this sense of purpose-lessness, which he had no difficulty in affirming:

> We are the children of chaos, and the deep structure of change is decay. At root, there is only corruption, and that unstemmable tide of chaos. Gone is purpose; all that is left is direction. This is the bleakness we have to accept as we peer deeply and dispassionately into the heart of the Universe.

All rather bleak, no doubt, but a perfectly legitimate angle on the second law of thermodynamics. I was perfectly prepared to accept this intellectually, although it was emotionally a little challenging.

Although I once shared that angle on things, I do so no longer. When I began to think of the world as created, my outlook changed entirely. Different perspectives were opened up for me. The stars, of course, remained as they were, but the way I viewed them altered radically. No longer were they harbingers of transience. They were now symbols of the wisdom and care of a God who knew and loved me. The words of Psalm 8 expressed my new attitude rather well:

> *When I look at your heavens, the work of your fingers,*
> *the moon and the stars that you have established;*
> *what are human beings that you are mindful of them,*
> *mortals that you care for them?*
> *Yet you have made them a little lower than God,*
> *and crowned them with glory and honor.*

The stars now became signs of the providence of God, who knows them and calls them by name (Psalm 147:4). No longer were the stars silent pointers to human transience; they were scintillating heralds of the love of

God. I was not alone in the universe, but walked and lived in the presence of a God who knew me and would never forget me. And the natural world was somehow "charged with the grandeur of God" (Gerard Manley Hopkins). And once nature is seen as God's creation, it can never be seen as ordinary again.

CHAPTER SEVEN

Disenchanting Nature: The Case of Richard Dawkins

"The dignity of the artist lies in his duty of keeping awake the sense of wonder in the world" (G. K. Chesterton). There can be no doubt that a sense of wonder is one of the most decisive motivations for the poet, the artist, the theologian, and the natural scientist. For the great philosophers of classical Athens, it led directly to philosophy—the human response to the glories of nature, resulting in both a correct understanding of the world and a good life within its bounds. Thomas Traherne (1637–74) wrote of the immense poetic inspiration he derived from a deep encounter with the natural world. The sight of green trees "transported and ravished me, their sweetness and unusual beauty made my heart to leap and almost mad with ecstasy, they were such strange and wonderful things."

A similar motivation underlies the best science, which springs from a love of nature and desire to deepen our appreciation of its mysteries. To master nature may be left to others of a more venial disposition; the sheer intellectual joy of science is what really matters. As Richard Powers puts it: "Science is not about control. It is about cultivating a perpetual sense of wonder in the face of something that forever grows one step richer and subtler than our latest theory about it. It is about reverence, not mastery." Even a casual reading of the enormous literary resources of the Christian tradition makes clear that this sense of wonder at nature is seen as being of fundamental religious importance. As Thomas Carlyle once noted, "worship is transcendent wonder."

For many—but not all—there is thus a natural link between scientific research and religious faith. Are not both about a sense of wonder and a longing to understand the world? It is important to note, however, that many natural scientists do not see any such link; some would dismiss such thoughts as fundamental lapses in human reasoning of positively Neanderthal proportions. The most interesting of this group is Richard Dawkins, professor of the public understanding of science at Oxford University. Dawkins merits further discussion, given the importance of religion for our theme.

Quackery: Richard Dawkins on Religion

Why, many of his readers wonder, is Dawkins so hostile to religion? In December 1991, the distinguished British novelist Fay Weldon contributed an article to the *Daily Telegraph* criticizing scientists for being arrogant and presumptuous at points, asking questions that did not really bother most people. Reacting against this article, which he styled a "hymn of hate" and a "rant against the scientists," Dawkins wondered "where such naked hostility comes from." I find myself asking the same question when reading Dawkins's works. The target may be different from Weldon's, but the tone of Dawkins's "hymn of hate" seems just as strident as that of his opponent. Why such naked hostility toward religion? Sure, some religious people do some very bad things, just like their antireligious opponents. Sure, they are arrogant at times. But this is hardly a vice that is limited to religious people, and it hardly merits the sustained invective of hatred (might we dare to call it a hymn of hate or a rant against religion?) that emanates from Dawkins's published writings and television appearances.

Dawkins often seems like the polemical inversion of a Bible-bashing fundamentalist preacher, peddling his own certainties, excoriating the views of his rivals, and mocking the mental and moral abilities of those foolish enough to disagree with him. There are many who reject religion for intellectual reasons, yet avoid the kind of antireligious animus that is so distinctive a feature of Dawkins's writings. In 1996, Dawkins was named Humanist of the Year. In his acceptance speech, published the following year in the journal *Humanist,* Dawkins set out his agenda for the eradication of what he regarded as the greatest evil of our age:

> It is fashionable to wax apocalyptic about the threat to humanity posed by the AIDS virus, "mad cow" disease, and many others, but I think a case can be made that *faith* is one of the world's great evils, comparable to the smallpox virus but harder to eradicate. Faith, being belief that isn't based on evidence, is the principal vice of any religion.

The views set out in this article are puzzling. Dawkins argues that religion leads to intolerance and hence to killing people. That there is merit in that criticism is obvious. Yet hatred and the consequent killing of individuals and communities are hardly the exclusive preserve of religion. The greatest atrocities of the twentieth century were committed by regimes that espoused atheism, often with a fanaticism that Dawkins seems to think is reserved only for religious people. A desire to eliminate religion at the intellectual level has the most unfortunate tendency to lead others to do this at the physical level. The firing squads that Stalin sent to liquidate the Buddhist monks of Mongolia—to name one obvious instance of an antireligious atrocity that can hardly be overlooked in a fair discussion of this point—gained at least something of their fanaticism and hatred of religion from those who told them that religion generated fanaticism and hatred, in addition to offering people a false consolation in life.

I cannot help but wonder why Dawkins seems to turn a blind eye to the awkward fact that the most appalling brutalities inflicted upon humanity during the twentieth century were committed by those who wanted to eradicate religion. It suits Dawkins's purposes well to maintain that only religious people commit massacres and other outrages. But in the real world—which I assume Dawkins regards as rather important in this matter—antireligious ideologies are just as brutal as their religious counterparts. I am also puzzled why Dawkins should speak of the "eradication" of faith when Adolf Hitler employed much the same way of speaking when seeking to eradicate one specific faith community—the Jews. Dawkins, I am sure, would condemn genocide as vigorously as anyone; yet his rhetoric of the "eradication" of faith resonates uncomfortably with both Stalinist and Nazi agendas.

In the Renaissance, the term "humanism" was used to refer to the quest for eloquence of expression and artistic excellence. It sought to foster the artistic and cultural advancement of humanity and gave rise to some of the greatest religious works of art that Western culture has ever known. It is a matter for some sadness that "humanism" has degenerated into this antireligious diatribe. It seems seriously out of place in our era, which places great emphasis upon respecting individuals, communities, and their beliefs, and tolerating differences. Quite frankly, Dawkins's strident antireligious advocacy seems to belong to another era. I respect his criticisms of religion; yet it is not too much to hope that he might learn to live with and respect people of religious convictions—including, it may be stressed, some 40 percent of the scientific community—instead of demanding the eradication of their faiths.

An Alternative Response: Freeman Dyson

It is instructive to compare Dawkins's intolerant rhetoric of dismissal with the more informed, considered, and judicious comments of Freeman Dyson, professor of physics at the Institute for Advanced Studies, Princeton. In an address on the relation of science and religion, delivered in Washington on May 16, 2000, Dyson shows that he is a much more historically informed, as well as personally tolerant, scientist than Dawkins. Dyson knows about the Stalinist purges and the coercive enforcement of atheism. Where Dawkins rehashes the stale and discredited Enlightenment myth of atheism as a liberator, Dyson is aware of the more sinister realities of history:

> We all know that religion has been historically, and still is today, a cause of great evil as well as great good in human affairs. We have seen terrible wars and terrible persecutions conducted in the name of religion. We have also seen large numbers of people inspired by religion to lives of heroic virtue, bringing education and medical care to the poor, helping to abolish slavery and spread peace among nations . . . The two individuals who epitomized the evils of our century, Adolf Hitler and Joseph Stalin, were both avowed atheists. Religion cannot be held responsible for their atrocities. And the three individuals who epitomized the good, Mahatma Gandhi, Martin Luther King and Mother Teresa, were all in their different ways religious.

Dyson makes the point that both religious and scientific people can go far beyond their competencies in making extravagant and immodest statements, which bring their disciplines into disrepute within the community at large:

> Science and religion are two windows that people look through, trying to understand the big universe outside, trying to understand why we are here. The two windows give different views, but they look out at the same universe. Both views are one-sided, neither is complete. Both leave out essential features of the real world. And both are worthy of respect. Trouble arises when either science or religion claims universal jurisdiction, when either religious dogma or scientific dogma claims to be infallible. Religious creationists and scientific materialists are equally dogmatic and insensitive. By their arrogance they bring both science and religion into disrepute. The media exaggerate their numbers and importance. The media rarely mention the fact that the great majority of religious people belong to moderate denominations that treat science with respect,

or the fact that the great majority of scientists treat religion with respect so long as religion does not claim jurisdiction over scientific questions.

Without wishing to endorse any particular model of the interaction of science and religion, including the attractive and gracious proposal that Dyson himself advocates, one cannot help but be impressed by his openness to the issues.

Dawkins, in contrast, seems less than interested in being fair to Christianity in particular or religion in general. His representation of Christianity is singularly less than accurate. For example, consider the following characterization of faith, which we find in a lecture given by Dawkins at the Edinburgh International Science Festival in 1992: "Faith is the great cop-out, the great excuse to evade the need to think and evaluate evidence. Faith is belief in spite of, even perhaps because of, the lack of evidence . . . Faith is not allowed to justify itself by argument." This is a bizarre misrepresentation of the Christian position, which I find deeply disappointing. Presumably, this is linked with Dawkins's frequently restated view that the persistence of religion owes much to the gullibility of young people who will believe anything they are told in their early years. Dawkins's caricature of Christianity may well carry weight with his increasingly religiously illiterate or religiously alienated audiences, who find in his writings ample confirmation of their prejudices, but merely persuades those familiar with religious traditions to conclude that Dawkins has no interest in understanding what he critiques.

Faith, Reason, and Science: A Response to Dawkins

I doubt if Dawkins will have very much interest in the matter, but for the benefit of readers who are disinclined to accept Dawkins's caricatures of Christian beliefs, the following may be instructive. The classic Christian tradition has always valued rationality and does not hold that faith involves the abandonment of reason or the absence of evidence. Indeed, the Chris-

tian tradition is so strong on this matter that it is often difficult to understand where Dawkins got these ideas. I have in front of me some of the
works of three leading Christian philosophers—Richard Swinburne (Oxford University), Nicolas Wolterstorff (Yale University), and Alvin Planti-
nga (University of Notre Dame). Without exception, their writings focus
on how one can make "warranted" or "coherent" statements concerning
God. It is a telling mark of our times that Dawkins's readers appear to assume that, because he knows his molecular biology, he is immediately qual-
ified to pontificate with competence on other matters. It is always nice to
think that there are exceptions to what G. M. Young, the noted scholar of
Victorian England, so brilliantly caricatures as "The Waste Land of Ex-
perts, each knowing so much about so little that he can neither be contra-
dicted nor is worth contradicting." Yet I cannot help but feel that Dawkins
has not really studied what he so easily dismisses. Dawkins is of the view
that religious faith is "the great excuse to evade the need to think and evaluate evidence." I have no doubt that there are some religious people who are
mentally lazy and do not think very much about their faith. Yet Dawkins's
writings lead me to think this intellectual laziness or complacency is not
limited to religious communities, but may be generously represented among
its critics as well.

Two areas of Christian theology have addressed this issue of the basis
of faith for the best part of two thousand years and have generated a substantial literature which Dawkins might care to read sometime. The disciplines of "natural theology" and "apologetics" both set out to ask what can
be known of God through the natural order, by intelligent and rational inquiry. It is most emphatically *not* based upon an evasion of evidence. Nor
does it inculcate ignorance and deception as core values, as Dawkins appears to think.

Let's take a simple example—Thomas Aquinas's arguments for God's
existence, which occur quite close to the beginning of his massive work
Summa Theologiae (The Totality of Theology), written during the thirteenth
century. This opens with a sustained discussion of the relation between
faith and reason, including the question of the evidential foundations of
faith. The point being made is that a careful critical consideration of the

nature of the world, particularly its ordering and structuring, lends weight to belief in the existence of God. Aquinas does not actually claim that such considerations prove this existence; the argument is much more along the lines of "providing support for" or "resonating with" faith rather than being its epistemological foundation. But the point being made is clear: intelligent reflection on the nature of the world is part of the case for faith. There is no question of a "cop-out," to use Dawkins's dismissive turn of phrase. We are speaking about a serious and committed reflection on the nature of the world, leading to the development of responsible explanations of its nature. It is no accident that 40 percent of natural scientists profess active religious belief; they merely bring to their faith the same commitment to truth, love of nature, and longing for understanding that underlie the natural sciences. Or, as some of them would argue—here inverting my comments—they merely bring to their scientific research and teaching the same commitment to truth, love of nature, and longing for understanding that underlies their faith.

There has been no shortage of historians who have argued that the development of the natural sciences in western Europe rests on precisely such a linkage between faith and evidence. While I personally am not persuaded that the historical evidence is quite as clear-cut as others suggest, it is important to note the case that is regularly made in the literature concerning the religious motivation for scientific research in the early modern period, and to note that there is ample—if not conclusive—evidence of a generally positive relation between science and faith throughout the early modern period.

This can be illustrated from the writings of John Calvin, the leading Protestant theologian of the sixteenth century. Calvin positively encouraged the scientific study of nature through his stress upon the orderliness of creation. He argued that both the physical world and the human body testify to the wisdom and character of God:

> In order that no one might be excluded from the means of obtaining happiness, God has been pleased, not only to place in our minds the seeds of religion of which we have already spoken, but

to make known his perfection in the whole structure of the universe, and daily place himself in our view, in such a manner that we cannot open our eyes without being compelled to observe him ... To prove his remarkable wisdom, both the heavens and the earth present us with countless proofs—not just those more advanced proofs which astronomy, medicine and all the other natural sciences are designed to illustrate, but proofs which force themselves on the attention of the most illiterate peasant, who cannot open his eyes without seeing them.

Calvin thus commends the study of both astronomy and medicine. They are able to probe more deeply than theology into the natural world and uncover further evidence of the orderliness of the creation and the wisdom of its creator.

It may be argued that Calvin gave a new religious motivation to the scientific investigation of nature. This was now seen as a means of discerning the wise hand of God in creation and thus enhancing both belief in his existence and the respect in which he was held. The Belgic Confession (1561), a Calvinist statement of faith which exercised particular influence in the Lowlands (an area that would become noted for its botanists and physicists), declared that nature is "before our eyes as a most beautiful book in which all created things, whether great or small, are as letters showing the invisible things of God to us." God can be discerned through the detailed study of his creation.

These ideas were taken up with enthusiasm within the Royal Society, the most significant organization devoted to the advancement of scientific research and learning in England. Many of its early members were admirers of Calvin, familiar with his writings and their potential relevance to their fields of study. Thus Richard Bentley delivered a series of lectures in 1692, based on Newton's *Principia Mathematica* (1687), in which the regularity of the universe, as established by Newton, is interpreted as evidence of the wisdom of God. There are unambiguous hints here of Calvin's reference to the universe as a "theatre of the glory of God," in which humans are an appreciative audience. The detailed study of the creation thus leads to an increased awareness of the wisdom of its creator. A fundamental religious motivation is generated for the scientific investigation of nature.

A further point concerns the Christian view that there is a basic correlation between human rationality and the rationality of the natural order, which is grounded in the doctrine of creation. The human mind seems able to grasp and partly replicate the structures of the world. As Edward Young put it in his *Night Thoughts* (1742–45), the human senses

> . . . *take in at once the landscape of the world,*
> *At a small inlet, which a grain might close,*
> *And half create the wondrous world they see.*

The Cambridge theoretical physicist and theologian John Polkinghorne points to the importance of this issue, noting the need for offering an explanation of why the human mind is able to uncover and grasp the structures of the world:

> We are so familiar with the fact that we can understand the world that most of the time we take it for granted. It is what makes science possible. Yet it could have been otherwise. The universe might have been a disorderly chaos rather than an orderly cosmos. Or it might have had a rationality which was inaccessible to us . . . There is a congruence between our minds and the universe, between the rationality experienced within and the rationality observed without. This extends not only to the mathematical articulation of fundamental theory but also to all those tacit acts of judgement, exercised with intuitive skill, which are equally indispensable to the scientific endeavour.

That human beings have been remarkably successful in investigating and grasping something of the structure and workings of the world is beyond dispute. Precisely why the rationality of the world should be so accessible to human beings remains more puzzling. Polkinghorne offers a Christian explanation of this phenomenon:

If the deep-seated congruence of the rationality present in our minds with the rationality present in the world is to find a true explanation, it must surely lie in some more profound reason which is the ground of both. Such a reason would be provided by the Rationality of the Creator.

The basic Christian idea that humanity is created in the "image of God" has long been seen by Christian theologians as offering both an explanation of the human capacity to understand the world and a stimulus to a greater encounter and engagement with the natural order. While this idea can be found throughout Christian history, it is stated with particular clarity by Augustine of Hippo in the early fifth century: "The image of the creator is to be found in the rational or intellectual soul of humanity . . . [which] has been created according to the image of God in order that it may use reason and intellect in order to apprehend and behold God." This basic idea lies behind the Christian engagement with the natural world, especially in the sixteenth and early seventeenth centuries. Thus the astronomer Johann Kepler, who made huge advances in our understanding of planetary orbits, had no doubt that the reason for the success of mathematics in clarifying the nature of these orbits lay in the creation of the world and the human mind by God:

> In that geometry is part of the divine mind from the origins of time, even from before the origins of time (for what is there in God that is not also from God?) it has provided God with the patterns for the creation of the world, and has been transferred to humanity with the image of God.

A similar point was made by Galileo, who attributed the success of his astronomical theories to mathematics being grounded in the being of God.

According to Dawkins's reading of history, of course, religious people should have been implacably hostile to the sciences. For religious people to like the sciences is about as likely as turkeys enjoying Thanksgiving. The historical evidence simply does not permit such an extravagant conclusion to be drawn, although there has been no shortage of those who sought to do so. For example, the controversy between Galileo and the church authorities is often portrayed as a direct confrontation between science and religion, especially by those writers wishing to perpetuate the myth that science and religion are permanently at war. As close historical scrutiny of this episode has shown, however, the reality is quite different, and rather more interesting, involving the complexities of political patronage at a particularly unstable juncture in the history of the papal court, leaving Galileo on the losing side of a court intrigue. That, however, is another story.

Dawkins's Simplistic Take on Science

Richard Dawkins seems to inhabit a world that is neatly divided into two realms. One is the hard-nosed realm of the natural sciences, in which everything that is believed may be proved. The other, whether intentionally or otherwise, is a world of delusion, chicanery, deceit, and superstition—a religious world in which people believe certain irrational things in spite of the evidence against such beliefs. This division of the world into the kingdoms of light and darkness may suit Dawkins's rhetorical style. It also resonates reasonably well with the unstated (and outdated) philosophical positivism that seems to underlie his statements on the working methods of the natural sciences. There are, however, certain very obvious difficulties with what Dawkins proposes, and it is important to note what they are.

First, Dawkins sees the world in black and white, with no shades of gray. On the one hand, we have the sciences, which are the only valid means of gaining knowledge. On the other, we have religion, which is blatant deception and illusion. Dawkins is thus left devoid of any discriminatory apparatus to make distinctions between various religious, supernatural, paranormal, and simply bizarre beliefs. Dawkins seems to think that belief in God is enough to consign you to the ranks of the untouchables, marking you out as an intellectual pariah. He draws a line in the intellectual sand, defined by the assumptions of what seems to be an outdated scientific positivism typical of the late nineteenth century, but which is not taken with any great seriousness by the philosophers of science of the twenty-first century, who tend to regard this as an oddity of largely historical interest.

Having drawn this arbitrary and virtually indefensible line in the sand, Dawkins places to his left all those who limit their beliefs to those that can be proved by the natural sciences, and on his right those who hold beliefs, usually religious, that cannot be proved in this way. There is, of course, an astonishing variation within those beliefs. That does not seem to trouble Dawkins; indeed, he seems to suggest that the intellectual lunacy and moral degeneracy of some of the more extreme variants can be applied to them all, thus allowing him to insinuate an unconvincing plea of "guilt by association."

The weakness, not to mention the lack of philosophical rigor, of this position can be seen by comparing his simplistic analysis with the outlook of one of the greatest British natural scientists of all time—Michael Faraday (1791–1867). Faraday's scientific work can be argued to have laid the foundations of all subsequent technology based on the use of electricity, including the modern electric motor, generator, and transformer. Faraday was also widely regarded as the greatest scientific lecturer of his day, who did much to publicize the great advances of nineteenth-century science and technology through his articles, correspondence, and especially the Friday evening discourses he established at the Royal Institution. The Royal Institution Christmas lectures for children, also begun by Faraday, continue to this day.

Faraday was an active Christian and saw a natural connection between his Christian faith and his scientific research. Following a line of thought that was widely canvassed within British Christianity since Newton, Faraday argued that "every machine and contrivance suggests to us an artist or contriver"; this led him to affirm that the "book of nature"—that is, the empirical natural world—"sufficiently evinces its author." Faraday believed it perfectly acceptable to infer the existence of a creator God from the evidence of the natural world, although he was firmly of the view that little could be known of the nature of that God in this way.

Yet Faraday was quite clear that there were many religious and popular superstitions that could not be allowed the same degree of credibility. One incident in particular illustrates this point. In 1848, some reports of unusual events began to emanate from Rochester, in upstate New York. Two sisters reported communications from a spirit world in the form of rappings. It was not long before much of America and subsequently Britain found itself caught up in a craze for mediums and séances, as a new fascination in spiritualism swept through the drawing rooms of the middle classes and beyond.

On June 30, 1853, Faraday published a letter in the London *Times* reporting his conviction that these alleged spiritualist manifestations were crude deceptions. This was not based on prejudice, but on his own experimental observations which led him to conclude that they were little more than blatant fabrications. He called for a reappraisal of the British education system, which failed to teach enough about the natural sciences and thus allowed the British public to fall victim to such credulity through the lack of any critical faculties.

Faraday illustrates the weakness of Dawkins's oversimplifications. Dawkins's rhetoric on occasion seems to run away with him and leads him to suggest—or at least to fail to deny what seems implicit in his writings—that there is a rigid line to be drawn between sanity and lunacy, between proven beliefs and irrational superstition, and that only those committed to a positivist—which Dawkins equates with a scientific—view of knowledge are on the correct side of that line.

Second, it must be questioned whether science "proves" things with quite the certainty that Dawkins allows. As Roger Bacon points out in many of his writings, the essence of the natural sciences is to proceed by induction, inferring the existence of certain laws or hypothetical entities through what is observed. Often, their existence cannot be proved, simply because the necessary experimental evidence lies beyond the technology currently available. "Anomalies" often exist alongside theories that indicate that these should not take place. Only naive falsificationists seriously believe that evidence that runs counter to a theory demands its immediate refutation. The question is whether that anomaly will ultimately be explained within the existing theory through an advance in understanding, or whether it will ultimately prove to be a decisive consideration in the rejection of that theory.

Physics provides us with an important case study illustrating this point rather neatly. For about three quarters of a century, physicists have had to live with the tension generated by the awkward fact that two of its foundational theories—the theory of general relativity and quantum mechanics—are less than perfectly reconciled with each other, perhaps even to the point of conflicting at certain key points. A naive approach to scientific theorizing would demand that one or the other be abandoned, as lacking in warrant. How could you hope to "prove" such theories when they are in conflict like this? Yet the wiser view, which predominates the philosophy and practice of the sciences, is that such tensions are to be expected. It is quite common for theory and observation to exist in a less-than-perfect state of correlation, due to the imperfections of our understanding. Such tensions may lead to an increase in scientific study of certain theories, but they most emphatically do *not* demand their abandonment on account of a degree of mismatch between theory and observation.

One thus cannot really speak of "proving" theories, unless this word is taken in a significantly reduced sense, meaning something like "having reason to believe that this is the best possible explanation, but being aware that there are others that cannot always be excluded." An important theoretical issue at this point is the "underdetermination of theory by evidence," a well-attested difficulty in relation to the verification of theories. In brief, this difficulty is that a given body of experimental evidence can generally be interpreted in a number of ways, with closure often being difficult to secure. "Proof" is thus a remarkably elusive concept in the real scientific world, however much it may be bandied about by the publicists of scientific imperialism.

Dawkins's hero Charles Darwin (1809–82) was well aware of the coexistence of theory and anomaly. A commendable personal and intellectual modesty graces his works, from which his successors might learn something. Although the historical account of how Darwin arrived at his theory has perhaps been subject to romantic embellishment, it is fairly clear that the driving force behind his reflections was the belief that the morphological and geographical phenomena could be convincingly accounted for by a single theory of natural selection. Darwin himself was quite clear that his explanation of the biological evidence was not the only one that could be adduced. He did, however, believe that it possessed greater explanatory power than its rivals.

In the end, Darwin recognized that his theory had many weaknesses and loose ends. For example, it required that speciation (one species developing into two separate species at one geographic location) should take place; yet the evidence for this was conspicuously absent. Darwin himself devoted a large section of the *Origin of Species* to detailing difficulties with his theory, noting in particular the "imperfection of the geological record," which gave little indication of the existence of intermediate species, and the "extreme perfection and complication" of certain individual organs, such as the eye. Nevertheless, he was convinced that these were difficulties or anomalies that could be tolerated on account of the explanatory superiority of his approach. While Darwin did not believe that he had adequately dealt with all the problems that required resolution, he was still confident that his explanation was the best available, as the following passage from the end of the *Origin* makes clear:

> A crowd of difficulties will have occurred to the reader. Some of them are so grave that to this day I can never reflect on them without being staggered; but, to the best of my judgement, the greater number are only apparent, and those that are real are not, I think, fatal to my theory.

The reader of these words cannot help but be impressed by Darwin's personal honesty and intellectual integrity. Darwin *knew* the limits of his theory and of the scientific method in general. Some of his successors might do well to learn from his personal demeanor and intellectual humility, and not just his scientific ideas.

Dawkins seems to get carried away with his antireligious rhetoric by sliding from "this cannot be proved" to "this is false" with an alarming ease, apparently unaware of the lapses in reasoning along the way. Consider, for example, his comment in a debate on "Whether Science Is Killing the Soul" in response to a question from the audience concerning whether science can offer consolation in the way that religion can—for example, after the death of a close friend or relative: "The fact that religion may console you doesn't of course make it true. It's a moot point whether one wishes to be consoled by a falsehood." Dawkins slides effortlessly from "consolation does not make religion true" to "religion is false." Now, perhaps this is an entirely natural inference for Dawkins himself, given his deeply ingrained antireligious feelings. But it is not a logically valid transition. To set this ridiculous piece of rhetoric out more clearly, so that this point can be seen:

1.   A cannot be proved;
2.   Therefore A is false.

This is flaky logic at its worst: (2) simply does not follow from (1). I am sure that Dawkins knows perfectly well that the fact that something cannot be proved does not make it false. This represents a classic example of the elementary "epistemic fallacy," which makes ontology (the way things actually are) dependent upon epistemology (what can be known about things).

Dawkins also fails to take due note of the fact that today's allegedly proved scientific theories can become tomorrow's discarded theories. What was once regarded as the cutting edge of scientific advance is discredited, and set aside as an outmoded and outdated approach. Dawkins's approach does not take account of the historical erosion of what were once thought to be the proven results of the natural sciences—a theme that is of major importance in the history of science and that underlies the notion of "paradigm shifts" introduced by Thomas Kuhn. What was thought to be "proven" in 1890, for example, might be rejected by a later generation, in the light of further research.

As philosopher of science Karl Popper often pointed out, part of the paradox of the scientific method is that while science is the most critically tested and evaluated form of knowledge available, it is nevertheless tentative and provisional. Science is to be seen as an "unended quest," whose findings may be up-to-date but are never final. As techniques are refined and conceptual frameworks modified, the understandings of one generation of natural scientists give way to those of another. Although there is a clear degree of continuity between the understandings of successive generations, this is probably based more on the methods they applied than on the outcome of their application.

The history of the scientific enterprise makes it abundantly clear that theories that were widely accepted in one generation—on the basis of the best evidence available—were superseded in following eras. For example, Newtonian mechanics and his theory of universal gravitation were widely regarded as correct in the eighteenth and nineteenth centuries, not on account of any sociological factors predisposing the scientific community to accept them, but simply because they offered the best explanation of the available observations. Yet the Newtonian worldview has now been superseded by the Einsteinian, on account of the explanatory and predictive successes of the general theory of relativity. Perhaps the most famous example of this overconfidence in existing scientific paradigms is Lord Kelvin's assertion, made in the closing years of the nineteenth century, that all that remained for physics to achieve was to fill in the next decimal place. The abandonment of the concept of "ether" and advent of quantum mechanics would demonstrate the provisionality of much of what appeared to be settled at the end of that century.

Dawkins seems to overlook this awkward fact. He appears to treat the matter as settled by the mantra "science proves." Let us agree that the scientific method seeks to ascertain incontrovertible grounds for believing that things are true; in other words, it values experimental demonstration of its hypotheses, actively seeks this out, and will not take theories seriously unless there are good evidential grounds for doing so. But it is a long way—a *very* long way—from this to the idea that science *proves* everything. There are many scientific theories that could only be "proved" on grounds that ultimately lie beyond the scientific method. Interestingly, Popper himself believed that Darwin's theory of natural selection falls into this category, on account of the need to be able to demonstrate what happened in history.

It should also be noted that Dawkins's strongly reductionist approach to the natural sciences in general, and the question of valid human knowledge in particular, fails to take account of the human longing for the transcendent, and the deep sense that there is more to this world than can be contained in any philosophy, classic or modern. At one level, this is to be seen in the interest in multiperspectival approaches to reality; at another, in the widespread reaction against one particular reading of the natural sciences—namely, that there is nothing to life and knowledge other than what can be proved by the natural sciences.

As we have stressed throughout this book, this is not a necessary consequence of the scientific method, and it would be vigorously contested by many natural scientists, especially those with a good knowledge of the history and philosophy of the discipline. It is distressing that such an epistemologically imperialist account of the scope of the natural sciences has been allowed to gain such an influence within the scientific community, not least because of the damaging perception of arrogance that has arisen within the public understanding of science.

As E. F. Schumacher pointed out with great sadness in his last book, *A Guide for the Perplexed,* the view that the natural sciences determine *all* things declares many of the most fundamental questions of human longing to be illegitimate and the answers sincerely given to be totally false: "The maps produced by modern materialistic Scientism leave all the questions that really matter unanswered: more than that, they deny the validity of the questions." Yet a love for the natural sciences need not lead to this materialist reductionism, nor to the suppression of the human yearning for a transcendent dimension to life. In the final chapter of this work, we shall consider how a synthesis of the sciences and the Christian faith may be achieved, opening the way to fresh ways of viewing and understanding nature.

CHAPTER EIGHT

Reenchanting Nature: Dawkins, Keats, and the Rainbow

The natural sciences owe both their brilliant explanatory successes and their distinctive limitations to their methodology, which is grounded on sensorially perceived reality. The development of new technology has increased the capacity of the human senses considerably. The telescope allows us to observe the rings of Saturn and moons of Jupiter and the microscope the fine structure of bacteria—things that the unaided human eye has difficulty in seeing. At other times, we can indirectly observe things that were once hidden—as when a subatomic particle leaves a distinctive visible trail in a cloud chamber. This observation-driven methodology of the natural sciences has ensured its successes and triumphs in areas where this method can be applied. Yet it needs to be made clear that there are areas in which this method simply cannot be applied. If natural scientists choose to make comment on these areas, they do so from a standpoint that lacks the rigorous experimental grounding of its proper areas of competence. The classic example of this is the question of God.

Strictly speaking, the natural sciences have nothing to say concerning the existence or otherwise of God. The observation-driven agenda of the natural sciences is constrained, by its very methodology, to deal only with sensorially perceived reality. Let us readily grant that it deals with these observables remarkably well and that it fully deserves its reputation for competence when this methodology may be applied. Yet the natural sciences simply cannot comment with competence on matters that cannot be observed or detected, directly or indirectly. "It cannot be observed" can never legitimately be restated as "it does not exist." The distinctive character of the sciences thus means that they can say nothing about the existence or otherwise of a transcendent reality, beyond interesting but ultimately speculative assertions or conjectures—for example, that it is just another interesting invention of the human mind resulting from certain neural configurations.

Sadly, there have been some natural scientists who would have us believe that their views on the meaning of life or the existence of God are to be regarded as just as reliable as their views on the value of Hubble's constant or the refractive index of water. The temptation to transfer authority from the scientist or the sciences in their own spheres of competency to other areas has always been present, and is to be firmly resisted. This is especially the case with Richard Dawkins's strongly reductionist approach to reality, best seen in his *Unweaving the Rainbow* (1998).

Dawkins on Unweaving the Rainbow

In his *Unweaving the Rainbow*, Dawkins takes issue with the views set out in Keats's 1820 poem "Lamia," which, as we saw in an earlier chapter, expresses the concern that reducing nature to scientific theories empties nature of its beauty and mystery and reduces it to something cold and clinical:

> *Do not all charms fly*
> *At the mere touch of cold philosophy?*
> *There was an awful rainbow once in heaven:*
> *We know her woof, her texture; she is given*

In the dull catalogue of common things.
Philosophy will clip an Angel's wings.

Dawkins regards this kind of stuff as typical antiscientific nonsense, which rests on the flimsiest of foundations. A good dose of scientific thinking would have sorted Keats out in no time:

> Why, in Keats' "Lamia," is the philosophy of rule and line "cold," and why do all charms flee before it? What is so threatening about reason? Mysteries do not lose their poetry when solved. Quite the contrary; the solution often turns out more beautiful than the puzzle and, in any case, when you have solved one mystery you uncover others, perhaps to inspire greater poetry.

Dawkins illustrates this point by drawing attention to the consequences of Newton's analysis of the rainbow: "Newton's dissection of the rainbow into light of different wavelengths led on to Maxwell's theory of electromagnetism and thence to Einstein's theory of special relativity."

The points that Dawkins makes are important and valid. Perhaps the road from Newton to Maxwell and thence to Einstein was rather more troublesome and complex than Dawkins's prose suggests, but the connection certainly exists. And if the unweaving of the rainbow led to the discovery of such greater mysteries (presumably perfectly capable of being expressed poetically, if poets could get their minds around the rather difficult ideas involved), then how can anybody suggest it was a foolish or improper thing to do?

We can see in Dawkins's writings an attitude toward poetry akin to that espoused by the British empiricist philosopher David Hume: "Poets themselves, though liars by profession, always endeavour to give an air of truth to their fictions." A similar attitude can be found in Peter Atkins's remarkable essay "The Limitless Power of Science":

> Although poets may aspire to understanding, their talents are more akin to entertaining self-deception. They may be able to emphasize

delights in the world, but they are deluded if they and their admirers believe that their identification of the delights and their use of poignant language are enough for comprehension. Philosophers too, I am afraid, have contributed to the understanding little more than poets . . . They have not contributed much that is novel until after novelty has been discovered by scientists.

For Dawkins, things are also admirably clear. Scientists tell the truth, occasionally in less-than-inspiring prose; poets, on the other hand, dislike and distrust science and generally know nothing about it. Dawkins believes that Keats argues something like this: if we know how the rainbow works, it will destroy its beauty, and we won't be able to appreciate it anymore. It would be like telling a small child that there is no Santa Claus. How silly! Anyone can see that the rainbow remains just as beautiful if we know how it works. In fact, we can appreciate its beauty to the full. Keats wrote these words while he was a young man. When he grew up, he might have become wiser and learned more about the sciences along the way. According to Dawkins, "Keats believed that Newton has destroyed all the poetry of the rainbow by reducing it to the prismatic colours. Keats could hardly have been more wrong."

Dawkins insists that scientists can easily improve upon Keats's poetry. This has puzzled some of Dawkins's critics, who reasonably point out that it does not follow that good science generates good poetry. Consider, for example, the following lines of the "Hymn to Science" of Mark Akenside (1721–70), who saw science as debunking bogus philosophies and theologies:

> Science! thou fair effusive ray
> From the great source of mental day,
> Free, generous, and refin'd!
> Descend with all thy treasures fraught,
> Illumine each bewilder'd thought,
> And bless my lab'ring mind.

But first with thy resistless light,
Disperse those phantoms from my sight,
Those mimic shades of thee;
The scholiast's learning, sophist's cant,
The visionary bigot's rant,
The monk's philosophy.

All worthy stuff, no doubt, but one can hardly call this preachy tirade great poetry.

Dawkins's refutation of Keats has understandably won many plaudits from his fellow scientists, who have welcomed his dismissal of critics who claim that science's at times tedious and plodding message robs nature of her beauty and inspiration. I regard Dawkins's response to Keats as being unassailable *if and only if* Keats's concern was to excoriate the scientific investigation of nature and take refuge in the safety of a premodern world. When Keats is read against the background of the Romantic movement, however, the critique he offers of the natural sciences begins to take on a quite different meaning. Far from refuting Keats, Dawkins in fact confirms precisely the fears that Keats expressed. Let me explain.

The key to Keats's concern lies in his reference to "clipping" an angel's wings. For Keats, as for the classical tradition in general, the natural world is a gateway to the realm of the transcendent. Human reason could grasp at least something of the real world, enabling the imagination to reflect on what it signified beyond itself. Keats (and the Romantic movement at large) prized the human imagination, seeing this as a faculty that allowed insights into the transcendent and sublime. Reason, in contrast, kept humanity firmly anchored to the ground and threatened to prevent it from discovering its deeper spiritual dimensions. Romanticism encouraged a deepened appreciation of the beauties of nature, in part through a general exaltation of emotion over reason and of the senses over intellect. "Vision or Imagination is a Representation of what Eternally Exists, Really and Unchangeably" (William Blake). The imagination led the human mind upward, to glimpse something of its eternal destiny. Reason, for the Romantics, barred access to this spiritual world. "The primary imagination I hold to be the living Power and prime Agent of all human Perception, and as a repetition in the finite mind of the eternal act of creation in the infinite I AM" (Samuel Taylor Coleridge). A similar theme can be found with C. S. Lewis, who once remarked that "while reason is the natural organ of truth, imagination is the organ of meaning."

Keats's strong views on the capacity of the human imagination to disclose both truth and beauty is set out in a series of letters he wrote on the subject in the years prior to the publication of "Lamia." For example, consider the following lines from Keats's letter to Benjamin Bailey, dated November 22, 1817:

> I am certain of nothing but of the holiness of the Heart's affections and the truth of Imagination—What the imagination seizes as Beauty must be truth—whether it existed before or not for I have the same Idea of all our Passions as of Love they are all in their sublime, creative of essential Beauty . . . The Imagination may be compared to Adam's dream—he awoke and found it truth. I am the more zealous in this affair, because I have never yet been able to perceive how any thing can be known for truth by consequitive reasoning . . . Adam's dream will do here and seems to be a conviction that Imagination and its empyreal reflection is the same as human Life and its spiritual repetition.

With this framework in mind, it is easy to see the point that Keats is making and why Dawkins's critique does not really meet the point at issue.

For Keats, a rainbow is meant to lift the human heart and imagination upward, intimating the existence of a world beyond the bounds of experience. For Dawkins, the rainbow remains firmly located within the world of human experience. It has no transcendent dimension. The fact that it can be explained in purely natural terms is taken to deny that it can have any significance as an indicator of transcendence. The angel that was, for Keats, meant to lift our thoughts heavenward has had its wings clipped; it can no longer do anything save mirror the world of earthly events and principles.

Dawkins's vigorous, dismissive, and highly polemic rejection of religion (or any human quest for the transcendent, for that matter) corresponds precisely to what Keats feared. Despite Dawkins's insistence to the contrary, Keats had no fundamental problems with scientific explanations of the rainbow. His criticisms were directed against those who insisted that this was *all* that there was to a rainbow—who denied that the rainbow could have any *symbolic* significance, both heightening the human yearning for a transcendent realm and hinting at means of its resolution. Dawkins's outright and premature dismissal of any such dimension to life entails a reductionist materialism that denies and eliminates any transcendent dimension to life as some kind of quackery, superstition, or confidence trick (to pick up on a few of the antireligious overstatements that grace his writings).

Dawkins would vigorously contest any suggestion that he is a reductionist. In a recent public debate with biologist Steven Pinker, Dawkins commented thus on the idea:

"Reductionism is one of those words that makes me want to reach for my revolver. It means nothing. Or rather it means a whole lot of different things, but the only thing anybody knows about it is that it's bad, you're supposed to disapprove of it."

How can you be a reductionist if you are telling the truth? Others inflate their ideas, going far beyond what can be proved by the sciences. The so-called reductionist merely cuts out the nonsense talked by religious and other unscientific people about the world. Dawkins insists that the sciences are able to tell us how the world actually is. The reductionist is simply telling it the way it is, whereas the inflationist brings in all kinds of superfluous ideas—such as metaphysics, God, and other such spooky things—which have no business to be there in the first place. They cannot be proved to be there; therefore they may be discarded. Sciences are about *proving* that certain things are there, or that certain things are true. What lies outside the scope of the scientific method, therefore, must be regarded as unreal or illusory. The rainbow is what the natural sciences disclose it to be—and nothing more. Keats improperly attaches all kinds of metaphysical ideas to this natural phenomenon, which can be swept away as idle superstition through the application of the scientific method.

It is a matter for profound regret that Dawkins makes no attempt to empathize with Keats—to try to understand the fear that Keats expresses and its wider resonance within Western culture. Keats reacted against materialism, which he feared would rob human life of its purpose and meaning. There seems to be an obtuseness on Dawkins's part here—a studied refusal to take Keats's concerns seriously, dismissing him as a muddled poet who just needed to take Physics 101 to get his weird ideas sorted out. Yet Keats articulates anxieties that countless intelligent and thoughtful people entertain about the natural sciences and where they are leading us. Dawkins is likely to alienate them still further with his dismissive and polemical attitude to their concerns.

Keats and Dawkins share the belief that nature ought to evoke wonder, but differ in their understandings of how this could and should come about and what it might imply. Religion—which seems to many to offer perhaps the best way of regaining a sense of wonder about nature—is completely written off by Dawkins. Keats is profoundly and utterly wrong. If a rainbow makes us think about eternal life, the mystery of God, or some such thing, then it has no business to do so. The rational scientific analysis of the rainbow—which makes perfect sense, once the concept of a refractive index is grasped—is all that needs to be said. Anything else is pure superstition or imagination. What the natural sciences disclose is all that there is to be disclosed. Yet this viewpoint is itself deeply troublesome and has been the subject of considerable controversy. Dawkins is committed to a viewpoint that is sometimes referred to as naturalism, and at other times as scientism. The basic theme is simple enough, and underlies Dawkins's excoriation of any who critique the natural sciences for any reason. But is it right?

A Critique of Scientific Naturalism

"Naturalism is polemically defined as repudiating the view that there exists or could exist any entities or events which lie, in principle, beyond the scope of scientific explanation" (Arthur C. Danto). This working definition of naturalism allows us to appreciate where Dawkins is coming from. If something can't be seen or described by the natural sciences, it is not real. The result is admirably simple. Belief in God or anything that might be thought to lie beyond the realm of nature is repudiated as lacking any scientific foundation.

It is a simple view of life, and it will satisfy many. What you see is what you get. There are, however, considerable difficulties with the view. As an example, let us take the views of Ernst Mach, the noted German physicist of the late nineteenth century. Mach argued that the natural sciences deal with whatever is immediately given by the senses—in other words, whatever we can detect with our senses. Science is thus the investigation of the "dependence of phenomena on one another." If you can't see it, hear it, or touch it, there is no reason to believe that whatever you are talking about exists in the first place. While Mach was prepared to allow the use of "auxiliary concepts" as bridges linking one observation with another, he insisted that they had no real existence. They were simply useful tools, "products of thought" which "exist only in our imagination and understanding."

This approach is not taken with great seriousness by most working scientists. Why not? Because the application of the experimental method yields results that force us to the conclusion that there exist entities we cannot yet see—and which, indeed, we may never be able to see—but whose existence seems demanded by the experimental evidence. Mach's positivism—that is, the belief that what you see is all that there is—precludes this position.

To illustrate this from Mach's own writings, we need look no further than his rejection of the existence of atoms. Mach took a strongly negative view of the atomic hypothesis, insisting that atoms were merely theoretical constructs. Reports from his laboratory during the 1890s tell of Mach stalking around grumpily, challenging anyone who was unwise enough to mention the word "atom." "Show me one!" was his standard response to anyone using the a-word. In many ways, Mach's response represents a highly commonsensical approach. Common sense aims to eliminate gibberish, garbage, and superstition.

This approach is virtually unworkable as a serious account of the methods and working assumptions of the natural sciences. As John Polk-inghorne points out in his judicious study *Reason and Reality*, neither the nat-ural sciences nor Christian theology holds that it is necessary to be able to "see" something in order to believe in it. The issue is how cogent an expla-nation is offered, on the base of warranted evidence.

> We habitually speak of entities which are not directly observable. No one has ever seen a gene (though there are X-ray photographs which, suitably interpreted, led Crick and Watson to the helical structure of DNA) or an electron (though there are tracks in bub-ble chambers which, suitably interpreted, indicate the existence of a particle of negative electric charge of about 4.8×10^{-10} esu and mass about 10^{-27} gm). No one has ever seen God (though there is the astonishing Christian claim that "the only Son, who is in the bosom of the Father, he has made him known" (John 1.18)).

As Mach's premature dismissal of the atomic hypothesis made clear, there is more to reality than what can be seen. Atoms really were there; they just could not be "seen" with the technology at Mach's disposal. The same is believed of genes, electrons, and God.

Modern philosophy of science, particularly the forms of critical real-ism that are currently ascendant within that discipline, insist that a rigorous distinction must be made between epistemological and ontological issues—that is, between how something can be known and whether something is ac-tually there. The philosophy of science lying behind Mach's view holds that if something cannot be perceived, it is not there. A critical realist position holds that something can be there without our being aware of it. It is not necessary for something to be seen before it exists. Roy Bhaskar, perhaps the most significant advocate of "critical realism," argues that much that is wrong with more old-fashioned philosophies of science results from a gross confusion of epistemology and ontology. Bhaskar designates these er-rors as the "epistemic fallacy," which rests on the false assumption that the structures of the world depend upon human observation. For Bhaskar, real-ity does not depend upon human observation to come into existence. It is already there; the question is how we discern it, coming to know what is already there, in advance of its being known. We may not be able to see it—but that does not permit us to conclude that it is not there.

What we can observe is governed by our location in history. Before the invention of the telescope, the moons of Jupiter could not be observed, their orbits determined, and their relevance for "celestial mechanics" understood. Yet those moons were there long before we noticed them. Before the advent of multistage launch vehicles in the 1960s, it was impossible to see the far side of the moon. It is therefore necessary, as philosopher of science Rom Harré points out, to note that there are many things that are "presently lying beyond experience but are nevertheless anticipated to be objects of experience at some point in the future."

This is the inevitable outcome of the inductive method of scientific investigation and theorizing, which may lead to the proposal of certain unobserved entities to explain the behavior of those that can be observed. A simple example of this is provided by movements of the planet Uranus, whose orbital characteristics proved to be anomalous in the light of the predictions of astronomical theory. It was, of course, possible that Newton's entire theory of planetary motion would need to be discarded, and replaced with something different. However, a simpler explanation lay to hand—that there was a hitherto unknown planet beyond Uranus, whose gravitational pull was influencing Uranus's orbit. The anomalous orbital parameters of Uranus could be explained by this hypothetical planet exercising a gravitational pull on Uranus. But this was just a hypothesis, an inference from experience. In due course, the hypothetical planet in question (Neptune) was discovered independently (but on the basis of the same calculations) by Adams and Leverrier in 1846. Mathematical analysis, followed by intense observation, led to the discovery of a planet.

Yet Neptune already existed and was known through its effects (the influence on the orbit of Uranus) before its existence was confirmed. Neptune had not been "seen"; its existence was, however, widely accepted, due to the coherent explanation that it offered for what was otherwise a puzzling series of observations.

So what is the relevance of such reflections for religion? To appreciate the point at issue, we may consider Darwin's *Origin of Species* and the many observations that lay behind it. In this important work, Darwin offered an *explanation* of reality that he believed to be the best way of making sense of the world. It allowed him to account for many otherwise puzzling features of the world, such as the distribution of species and the existence of vestigial organs. Darwin lacked the resources to prove his theory but believed that it made enough sense for him to hold it with confidence. One day, he was sure, anomalies would be resolved, loose ends tied up. By the very nature of his theory, experimental analysis was impossible. What he proposed seemed to him to be the best explanation. It could not be verified as it stood. This did not, as we have seen, prevent Darwin from trusting in his theory.

The Christian—or, indeed, anyone aware of the spiritual dimensions of nature—believes that belief in God as creator and redeemer offers a compelling and attractive explanation of the way the world is. It offers an explanation of many things—such as the ordering of nature and the ability of the human mind to understand and represent this. (We explored this in the previous chapter.) Still, there are anomalies, puzzles, and apparent contradictions. The issue of pain and suffering in the world remains something of a puzzle, and at times troubles Christians considerably. Yet they hold on to their faith, believing that its explanatory ability and coherence are sufficient to justify it and that the difficulty will one day be resolved.

The intellectual morphology of these positions is not dissimilar, each representing an example of what Gilbert Harmann has called "inference to the best explanation." The basic issue is discerning how best to make sense of a considerable body of material. The noted nineteenth-century philosopher of science William Whewell described this process in terms of "consilience of inductions." By this, he meant the process by which different aspects of human experience are brought together to give a coherent view of the world. Both Darwin and the Christian may be argued to bring together a number of pieces of evidence, drawn from different areas of experience, and respectively hold that natural selection or Christianity represents warranted explanations of that evidence.

This approach, of course, is in direct contradiction to the reductionism of Dawkins and others, such as E. O. Wilson, who seem content to act as cheerleaders for imperialist approaches to the sciences. Wilson broadly understands reductionism as the view that the central concepts that characterize macrolevel phenomena in fields such as psychology, religion, art, and ethics can be translated into microlevel concepts, such as those of genetics and molecular biology; and these in turn can be defined or understood in terms of the concepts of physics.

To judge from his important 1998 work *Consilience: The Unity of Knowledge*, Wilson has a genuine and strong interest in matters of ethics. He declares that "ethics is everything" and urges his readers to protect biological diversity and to "expand resources and improve the quality of life." These are doubtless entirely worthy goals. But what are the *grounds* of these noble sentiments? Why should we believe them to be right? Wilson's chapter on ethics and religion is by far the weakest in his book. In it, he argues against the existence of both God and transcendent values, yet seems unaware that this leaves him vulnerable to the charge of asserting moral values that are purely arbitrary or temporary human conventions that are dependent upon culture and historical location. How can morality have credibility in a world of mere fact, from which God, religion, and any form of transcendent values have been eliminated?

Though showing no signs of being aware of the fact, Wilson has simply smuggled in a prior belief system under the cover of legitimate scientific explanations. While insisting that "science is neither a philosophy nor a belief system," Wilson seems to require precisely such a philosophy or belief system in order to lend credibility to the moral values he proposes. His work suffers from the same weakness that critics have discerned in the equally bold writings of the French postmodern philosopher Michel Foucault, who also proposes dispensing with such archaic notions as absolute truth or absolute values in order to affirm human liberty and combat all forms of oppression. Yet Foucault seems unaware of the implicit absolute moral values that saturate his own writing and thinking—values that can be summed up as "oppression is always bad, and liberty is always good." It would be interesting to know where these confident moral assertions come from, and how they are to be defended, given that Foucault has stripped away the foundations on which they ultimately rest. Wilson wants to affirm certain ecological and personal moral values, but seems to have overlooked the fact that he seems to have sawn off the branch of the tree on which they depended.

The Reenchantment of Nature: Reclaiming a Lost World

The pre-Socratics—the earliest Greek philosophers—argued endlessly about how it was possible to know the true nature of the world. It was clear to them that the universe was rationally constructed and that it could therefore be understood through the right use of human reason and critical reflection. Human beings possessed, they thought, an intrinsic ability to make sense of the universe. Socrates took this highly important line of thought still further, arguing that there was a fundamental link between the way the universe was shaped and the best way for human beings to live. To reflect on the nature of the universe was to gain insights into the nature of the "good life"—the best and most authentic way of living. Reflecting on the clues provided in the structuring of the world leads to an understanding of our identity and destiny.

The same idea emerges in some of the writings of the Old Testament, especially those focusing on the theme of wisdom. The wisdom writings of the Old Testament insisted that sense could be made of every aspect of life—from the movements of the stars to the behavior of people. It was all a question of combining close observation of the world with the right explanatory framework. Observing the world opens the door to an understanding of the greater matters of life. The Book of Job compares the human quest for wisdom to the mining of precious metals and stones from deep within the earth. They lie hidden from human view and must be sought out. Like the pearl of great price, true wisdom is profoundly worth seeking and possessing. Respectful and careful reflection on the deep structuring of existence—from the behavior of people and animals to the patternings of the natural world—held the key to a proper understanding of the nature and destiny of humanity.

I have argued in this book for the continuation of this great classical tradition, which has been all but destroyed in the past two centuries, initially through the rise of the Enlightenment and then through the subtleties of postmodernity. C. S. Lewis, one of the most distinguished representatives of this tradition in the twentieth century, made a telling point that is of central importance to our thinking on the foundations of a positive environmental ethos. In his *The Abolition of Man,* he argues that our attitude toward the world must be grounded in the deep structures of nature—structures that we did not place there and did not invent, but that were there before us, and must shape our responses to nature:

> For the wise men of old, the cardinal problem had been how to conform the soul to reality, and the solution had been knowledge, self-discipline and virtue. For magic and applied science alike the problem is how to subdue reality to the wishes of men; the solution is a technique.

The fundamental impulse of modernity was to dominate—to understand in order to master, as a conquering army might learn the language and customs of those who stood in their way, in order to ensure their more effective subjugation. Reality was to be subdued. Postmodernity offered a different vision, arguing that there was nothing intrinsic in nature that demanded any particular attitude toward it. Our understanding and valuation of nature is a matter of free human choice and is not dictated by the nature of things itself. There was no meaning within nature other than what we invented; no transcendent significance other than what we chose to make of it; no reason to view it in any way other than as personal taste or transient social conventions moved us. Humanity was completely free to construct its own vision of nature and act accordingly.

The impact of these dominant intellectual systems of the past two hundred years has been obvious. We have lost touch with the world of nature and have constructed our own worlds in its place. The intellectual pretensions of the modern period have led many to decouple themselves from the natural order. An increasing distance from the land has led us to neglect its needs and lose the sense of connection and dependence of which our forebears were so conscious. While some have suggested that Christianity is somehow implicated in this, the truth is otherwise. It is the rise of urbanization and technology that has placed barriers between modern humanity and nature. This lack of experience of nature causes some to exploit it, and others to romanticize it, creating a nostalgic view of humanity's past relationship with the environment that cannot stand up to scholarly examination.

We have become dislocated from our past as well as from nature. Christians of the twenty-first century need to rediscover the wisdom of earlier generations, which lived out their faith in close contact with the earth. It is not difficult to find renderings of the Christian faith that take our relationship with the environment to be an integral aspect of our relationship with God, which prize the cultivation of responsible attitudes to nature as a central task of Christian discipleship. The monastic traditions of the Egyptian and Syrian deserts in the fourth century, the Celtic Christian approach to nature of the seventh and eighth centuries, the Franciscan love for the natural order of the thirteenth and fourteenth centuries—all can guide and stimulate us, as we try to reconnect ourselves to nature. These great renderings of the Christian faith were firmly rooted in the Bible and sensitive to the world around them—a world that was seen and respected as a means of sustaining human life in the present and of reminding and reassuring believers of their future destiny in a renewed creation.

The basic theme of this book is simple. It suggests that we reclaim the idea of nature as God's creation and act accordingly, bringing attitudes and actions into line with beliefs. We have been entrusted, corporately and individually, with the jewel of God's creation and given the responsibility of tending and nurturing it, before passing it on to others. We are like curators of a great art collection, who are accountable to posterity as well as to the present for our tending of its treasures. We must learn to appreciate and prize this entrustment, as perhaps the greatest privilege this earth can offer.

And more than this: we must see nature as a continual reminder and symbol of a future renewed creation, a world that we do not yet know but believe to lie over the horizons of our human existence. It is as if we are homesick for a lost Eden, longing for a fulfillment that we know lies ahead of us but have not yet found. The natural order, as God's richly signed creation, is thus our place of living and of hoping. As John Keats saw the ornamentation on a Greek urn as mirroring the great human longing for meaning, value, and beauty, we must learn to see the present beauty of nature as a sign and promise of the coming glory of God, its creator. We must remember that others who are yet to come also need to glimpse these flashes of glory in everyday things, and preserve it for their sakes.

Christianity thus affirms what many others believe to be the case: that humanity is not entitled to think of itself as "possessing" nature. It has been loaned to us. We are accountable for our stewardship of its resources. Yet it places these shared principles on a footing that goes beyond human convention, taste, or fashion, insisting that they are to be seen as grounded in the character and will of God. The creation mandate, dismissed by Enlightenment rationalism as outdated and outmoded, can now be taken with its full seriousness, not least because the ruinous legacy of the Enlightenment can no longer be ignored.

Perhaps the most singular and striking aspect of the Christian perspective on creation lies in its role as the intimater of transcendence, hinting at a world beyond our experience, whose borders we may tread yet not cross. Our sense of wonder at the beauty of nature heightens our awareness of our limits, while heightening our longing to enter into what it seems to promise. Nature is to be cherished and valued, not simply for what it *is* but for what it *foreshadows*—a new creation, renewing and bringing to perfection the tired and ruined world that we know and seek to tend throughout what we fear may be its final illness.

To reenchant nature is to accept and cherish its divine origins and signification, not least in what it implies for our own nature and ultimate destiny. We dwell in this world, cherishing it while knowing it to be a staging post to somewhere even more wonderful. But while we linger on this earth, we may savor its beauty and stand in awe of its majesty. C. S. Lewis made this point in his early writing *The Pilgrim's Regress*. We are, he commented, like the man who knows "the fits of a strange desire, which haunt him from his earliest years, for something that cannot be named; something which he can describe only as 'Not this,' 'Far farther,' or 'Yonder.' "

Nature, when rightly understood, points beyond itself, to a "yonder" we shall one day know and inhabit. Richard Dawkins and others would have us clip this angel's wings, so that such an upward trajectory of the human imagination would be declared out of bounds. For others, it is a clue to the meaning of life. The poet John Gillespie Magee (1921–41) saw a pilot soaring above the earth as an image of a deeper spiritual journey, from the things of earth to the mind of God.

I have slipped the surly bonds of earth
And danced the skies on laughter-silvered wings . . .
Put out my hand and touched the face of God.

To reenchant nature is not merely to gain a new respect for its integrity and well-being; it is to throw open the doors to a deeper level of existence.

Works Consulted

Adas, Michael. *Machines as the Measure of Men: Science, Technology, and Ideologies of Western Dominance*. Ithaca, N.Y.: Cornell University Press, 1989.

Adorno, Theodor W., and Max Horkheimer. *Dialectic of Enlightenment*. London: Verso Editions, 1979.

Atkins, P. W. *The Second Law*. New York: W. H. Freeman, 1984.

Atkins, Peter. "The Limitless Power of Science." In *Nature's Imagination: The Frontiers of Scientific Vision*, edited by John Cornwell, 122–32. Oxford: Oxford University Press, 1995.

Bahro, Rudolf. *Rückkehr: Die in-Welt Krise Als Ursprung der Weltzerstörung*. Frankfurt am Main: Horizonte Verlag, 1991.

Bainbridge, William S., and Rodney Stark. *The Future of Religion: Secularization, Revival, and Cult Formation*. Berkeley, Calif.: University of California Press, 1985.

Barbour, Ian G. *Ethics in an Age of Technology*. San Francisco: HarperSanFrancisco, 1993.

———. "Experiencing and Interpreting Nature in Science and Religion." *Zygon* 29 (1994): 457–87.

———. *Myths, Models and Paradigms: A Comparative Study in Science and Religion*. New York: Harper & Row, 1974.

Barr, James. "The Image of God in the Book of Genesis: A Study of Terminology." *Bulletin of the John Rylands Library* 51 (1968): 11–26.

Bate, Jonathan. *Romantic Ecology: Wordsworth and the Environmental Tradition*. London: Routledge, 1991.

Baur, John I. H. *A Mirror of Creation: 150 Years of American Nature Painting*. New York: Friends of American Art in Religion, 1980.

Baym, Nina. "From Metaphysics to Metaphor: The Image of Water in Emerson and Thoreau." *Studies in Romanticism* 5 (1966): 231–41.

Bennett, J. A. "The Mechanics' Philosophy and the Mechanical Philosophy." *History of Science* 24 (1986): 1–28.

Benton, Ted. "Why Are Sociologists Naturephobes?" In *After Postmodernism: An Introduction to Critical Realism*, edited by José López and Garry Potter, 133–45. London: Athlone Press, 2001.

Berkowitz, Peter. *Nietzsche: The Ethics of an Immoralist*. Cambridge, Mass.: Harvard University Press, 1995.

Bernal, J. D. *The World, the Flesh and the Devil: An Enquiry into the Future of the Three Enemies of the Rational Soul*. London: Cape, 1970.

Berry, R. J., ed. *The Care of Creation*. Leicester: Inter-Varsity Press, 2000.

Bewell, Alan. *Wordsworth and the Enlightenment: Nature, Man and Society in the Experimental Poetry*. New Haven, Conn.: Yale University Press, 1989.

Blair, Ann. *The Theater of Nature: Jean Bodin and Renaissance Science*. Princeton, N.J.: Princeton University Press, 1997.

Boas, Marie. "The Establishment of the Mechanical Philosophy." *Osiris* 10 (1962): 442–520.

Borgmann, Albert. "The Nature of Reality and the Reality of Nature." In *Reinventing Nature: Responses to Postmodern Deconstruction*, edited by Michael E. Soulé and Gary Lease, 31–46. Washington, D.C.: Island Press, 1995.

Bosse, Hans. *Marx, Weber, Troeltsch. Religionssoziologie und Marxistische Ideologiekritik*. Vol. Nr 2, *Gesellschaft und Theologie. Abteilung: Sozialwissenschaftliche Analysen*. München: Kaiser, 1971.

Bradley, Ian. *Celtic Christianity: Making Myths and Chasing Dreams*. Edinburgh: Edinburgh University Press, 1999.

———. *The Celtic Way*. London: Darton Longman & Todd, 1993.

Briggs, John. *Fractals: The Patterns of Chaos: Discovering a New Aesthetic of Art, Science, and Nature*. London: Thames & Hudson, 1992.

Briggs, John C. *Francis Bacon and the Rhetoric of Nature*. Cambridge, Mass.: Harvard University Press, 1989.

Brockliss, L. W. B. "Aristotle, Descartes and the New Science: Natural Philosophy at the University of Paris, 1600–1740." *Annals of Science* 38 (1981): 33–69.

Brooke, John, and Geoffrey Cantor. *Reconstructing Nature: The Engagement of Science and Religion*. Edinburgh: T. & T. Clarke, 1998.

Brooke, John H. *Telling the Story of Science and Religion: A Nuanced Account*. Cambridge: Cambridge University Press, 1991.

Brooke, John Hedley. *Science and Religion: Some Historical Perspectives*. Cambridge: Cambridge University Press, 1991.

Brotemarkle, Diane. *Imagination and Myths in John Keats's Poetry*. San Francisco: Mellen Research University Press, 1993.

Burtt, Edwin Arthur. *The Metaphysical Foundations of Modern Physical Science*. Garden City, N.Y.: Doubleday Anchor, 1954.

Burwick, Frederick, and Paul Douglass. *The Crisis in Modernism: Bergson and the Vitalist Controversy.* Cambridge: Cambridge University Press, 1992.

Byrne, Peter A. *Natural Religion and the Religion of Nature.* London: Routledge, 1989.

Carnell, Corbin Scott. *Bright Shadow of Reality: Spiritual Longing in C. S. Lewis.* Grand Rapids, Mich.: Eerdmans, 1999.

Cartwright, Nancy. *Nature's Capacities and Their Measurement.* Oxford: Clarendon Press, 1989.

Caton, Donald. *What a Blessing She Had Chloroform: The Medical and Social Response to the Pain of Childbirth from 1800 to the Present.* New Haven, Conn.: Yale University Press, 1999.

Chandrasekhar, S. *Truth and Beauty: Aesthetics and Motivations in Science.* Chicago: University of Chicago Press, 1990.

Chenu, M. D. *Nature, Man and Society in the Twelfth Century.* Chicago: University of Chicago Press, 1968.

Clavelin, Maurice. *The Natural Philosophy of Galileo: Essay on the Origins and Formation of Classical Mechanics.* Cambridge, Mass.: M.I.T. Press, 1974.

Cobb, John. *Is It Too Late? A Theology of Ecology.* Beverly Hills, Calif.: Bruce Books, 1972.

Cohen, Jeremy. *"Be Fertile and Increase, Fill the Earth and Master It": The Ancient and Medieval Career of a Biblical Text.* Ithaca, N.Y.: Cornell University Press, 1989.

Collingwood, R. G. *The Idea of Nature.* London and New York: Oxford University Press, 1945.

Cotterill, Rodney. *No Ghost in the Machine: Modern Science and the Brain, the Mind and the Soul.* London: Heinemann, 1990.

Cronon, William. *Uncommon Ground: Toward Reinventing Nature.* New York: W. W. Norton & Co., 1995.

Davies, Paul. *The Mind of God: Science and the Search for Ultimate Meaning.* London: Penguin, 1992.

Davis, Philip G. *Goddess Unmasked: The Rise of Neo-Pagan Feminist Spirituality.* Dallas, Tex.: Spence Publishing Co., 1998.

Dawkins, Richard. "A Scientist's Case against God." *Independent,* April 20, 1992.

———. *Unweaving the Rainbow: Science, Delusion and the Appetite for Wonder.* London: Penguin Books, 1998.

Devereux, Paul, John Steele, and David Kubrin. *Earthmind: Communicating with the Living World of Gaia.* Rochester, Vt.: Destiny Books, 1989.

De Waal, Esther. *Celtic Light: A Tradition Rediscovered.* London: Fount, 1997.

———. *A World Made Whole: The Rediscovery of the Celtic Tradition.* London: Collins, 1991.

DeWitt, Calvin B. "Ecology and Ethics: Relation of Religious Belief to Ecological Practice in the Biblical Tradition." *Biodiversity and Conservation* 4 (1995): 838–48.

———. *The Environment and the Christian: What Does the New Testament Say about the Environment?* Grand Rapids, Mich.: Baker Book House, 1991.

———. *The Just Stewardship of Land and Creation.* Grand Rapids, Mich.: Reformed Ecumenical Council, 1996.

DeWitt, Calvin B., and Ghillean T. Prance. *Missionary Earthkeeping.* Macon, Ga.: Mercer University Press, 1992.

Dirac, Paul. "The Evolution of the Physicist's Picture of Nature." *Scientific American* 208, no. 5 (1963): 45–53.

Draper, John William. *History of the Conflict between Religion and Science.* New York: Daniel Appleton, 1874.

Eagleton, Terry. *The Idea of Culture.* Oxford: Blackwell Publishing, 2000.

————. *The Ideology of the Aesthetic.* Oxford: Blackwell, 1990.

————. *The Illusions of Postmodernism.* Oxford: Blackwell, 1996.

Evernden, Neil. *The Social Creation of Nature.* Baltimore, Md.: Johns Hopkins University Press, 1992.

Fastenrath, Heinz. *Ein Abriss Atheistischer Grundpositionen: Feuerbach, Marx, Nietzsche, Sartre.* Stuttgart: Klett, 1993.

Field, Michael J., and Martin Golubitsky. *Symmetry in Chaos: A Search for Pattern in Mathematics, Art and Nature.* Oxford: Oxford University Press, 1995.

Fish, Stanley Eugene. *Surprised by Sin: The Reader in Paradise Lost.* 2nd ed. London: Macmillan, 1997.

Fisher, Philip. *Wonder, the Rainbow, and the Aesthetics of Rare Experiences.* Cambridge, Mass.: Harvard University Press, 1998.

Folse, H. *The Philosophy of Niels Bohr: The Framework of Complementarity.* Amsterdam: North Holland, 1985.

Force, James E. "The Breakdown of the Newtonian Synthesis of Science and Religion: Hume, Newton and the Royal Society." In *Essays on the Context, Nature and Influence of Isaac Newton's Theology,* edited by R. H. Popkin and J. E. Force, 143–63. Dordrecht: Kluwer Academic Publishers, 1990.

Fraser, Hilary. *Beauty and Belief: Aesthetics and Religion in Victorian Literature.* Cambridge: Cambridge University Press, 1986.

Furley, David J. *Cosmic Problems: Essays on Greek and Roman Philosophy of Nature.* Cambridge: Cambridge University Press, 1989.

Garber, Daniel, *Descartes' Metaphysical Physics.* Chicago: University of Chicago Press, 1992.

Gay, Peter. *A Godless Jew: Freud, Atheism, and the Making of Psychoanalysis.* New Haven, Conn.: Yale University Press, 1987.

Gell-Mann, Murray. *The Quark and the Jaguar: Adventures in the Simple and the Complex.* London: Abacus, 1995.

Gilkey, Langdon. *Maker of Heaven and Earth: The Christian Doctrine of Creation in the Light of Modern Knowledge.* Garden City, N.Y.: Doubleday, 1959.

Gilkey, Langdon Brown. *Nature, Reality, and the Sacred: The Nexus of Science and Religion, Theology and the Sciences.* Minneapolis: Fortress Press, 1993.

Gimpel, Jean. *The Medieval Machine: The Industrial Revolution of the Middle Ages.* New York: Holt Rinehart and Winston, 1976.

Glacken, Clarence J. *Traces on the Rhodean Shore: Nature and Culture in Western Thought from Ancient Times to the End of the Eighteenth Century.* Berkeley, Calif.: University of California Press, 1967.

Goodenough, Ursula. *The Sacred Depths of Nature.* New York: Oxford University Press, 1998.

Goodman, David E., and Michael R. Redclift. *Refashioning Nature: Food, Ecology and Culture.* New York: Routledge, 1991.

Gregory, Frederick. *Nature Lost? Natural Science and the German Theological Traditions of the Nineteenth Century.* Cambridge, Mass.: Harvard University Press, 1992.

Groueff, Stéphane. *Manhattan Project: The Untold Story of the Making of the Atomic Bomb.* Boston: Little Brown, 1967.

Guardini, Romano. *Letters from Lake Como: Explorations in Technology and the Human Race.* Grand Rapids, Mich.: Eerdmans, 1994.

Hahn, Roger. "Laplace and the Vanishing Role of God in the Physical Universe." In *The Analytic Spirit,* edited by Harry Woolf, 85–95. Ithaca, N.Y.: Cornell University Press, 1981.

Hall, Douglas John. *Imaging God: Dominion as Stewardship.* Grand Rapids, Mich.: Eerdmans, 1986.

Hansen, Vagn Lundsgaard. *Geometry in Nature.* Wellesley, Mass.: A. K. Peters, 1993.

Hanson, N. R. *Patterns of Discovery: An Inquiry into the Conceptual Foundations of Science.* Cambridge: Cambridge University Press, 1961.

Haraway, Donna J. *Simians, Cyborgs and Women: The Reinvention of Nature.* New York: Routledge, 1991.

Hartlieb, Elisabeth. *Natur Als Schöpfung: Studien zum Verhältnis von Naturbegriff und Schöpfungsverständnis bei Günter Altner, Sigurd M. Daecke, Hermann Dembowski und Christian Link.* Frankfurt am Main and Berlin: Peter Lang, 1996.

Haught, John. *The Promise of Nature.* New York: Paulist, 1993.

Honner, John. *The Description of Nature: Niels Bohr and the Philosophy of Quantum Physics.* Oxford: Clarendon Press, 1987.

Hughes, J. Donald. *Ecology in Ancient Civilizations.* Albuquerque, N.M.: University of New Mexico Press, 1975.

———. *North American Indian Ecology.* 2nd ed. [El Paso, Tex.]: Texas Western Press, 1996.

Hughes, Jonathan R. T. *Ecology and Historical Materialism.* Cambridge: Cambridge University Press, 2000.

Israel, Joachim. *Martin Buber: Dialogphilosophie in Theorie und Praxis.* Berlin: Duncker & Humblot, 1995.

Johnson, George. *Strange Beauty: Murray Gell-Mann and the Revolution in Twentieth-Century Physics.* London: Jonathan Cape, 2000.

Johnson, Peter. *The Constants of Nature: A Realist Account.* Aldershot, Hants: Ashgate Publishing Ltd., 1997.

Jones, William Frank. *Nature and Natural Science: The Philosophy of Frederick J. E. Woodbridge.* Buffalo, N.Y.: Prometheus Books, 1983.

Jugurtha, Lillie. *Keats and Nature.* New York: Peter Lang, 1985.

Jungnickel, C., and R. McCormmach. *Intellectual Mastery of Nature: Theoretical Physics from Ohm to Einstein.* 2 vols. Chicago: University of Chicago Press, 1986.

Kellert, Stephen R. "Concepts of Nature East and West." In *Reinventing Nature: Responses to Postmodern Deconstruction,* edited by Michael E. Soulé and Gary Lease, 103–22. Washington, D.C.: Island Press, 1995.

Kerr, Fergus. *Immortal Longings: Versions of Transcending Humanity*. London: SPCK, 1997.

Koestler, Arthur. *The Ghost in the Machine*. London: Hutchinson, 1967.

Küng, Hans. *Freud and the Problem of God*. New Haven, Conn.: Yale University Press, 1979.

Larson, James L. *Interpreting Nature: The Science of Living from Linnaeus to Kant*. Baltimore, Md.: Johns Hopkins Press, 1994.

Lease, Gary. "Nature under Fire." In *Reinventing Nature: Responses to Postmodern Deconstruction*, edited by Michael E. Soulé and Gary Lease, 3–16. Washington, D.C.: Island Press, 1995.

Lecourt, Dominique. *Prométhée, Faust, Frankenstein: Fondements Imaginaires de l'Éthique*. Paris: Synthelabo, 1996.

Lenoble, Robert. *Mersenne ou la Naissance du Mécanisme*. Paris: Librarie J. Vrin, 1943.

Lewis, C. S. *The Abolition of Man*. London: Collins, 1978.

———. *The Pilgrim's Regress: An Allegorical Typology for Christianity, Reason and Romanticism*. London: Sheed & Ward, 1933.

Lewis, Linda M. *The Promethean Politics of Milton, Blake and Shelley*. London: University of Missouri Press, 1992.

Lindberg, David C., and Ronald L. Numbers. *God and Nature: Historical Essays on the Encounter between Christianity and Science*. Berkeley, Calif.: University of California Press, 1986.

Lloyd, Alan B., ed. *What Is a God? Studies in the Nature of Greek Divinity*. London: Duckworth, 1997.

Lloyd, Geoffrey E. R. "Greek Antiquity: The Invention of Nature." In *The Concept of Nature*, edited by John Torrance, 1–24. Oxford: Oxford University Press, 1992.

Longino, Helen E. "Toward an Epistemology for Biological Pluralism." In *Biology and Epistemology*, edited by Richard Creath and Jane Maienschein, 261–86. Cambridge: Cambridge University Press, 2000.

Lovelock, James. *The Ages of Gaia: A Biography of Our Living Earth*. New York: Norton, 1988.

———. *Homage to Gaia: The Life of an Independent Scientist*. New York: Oxford University Press, 2000.

Lupak, Mario John. *Byron as a Poet of Nature: The Search for Paradise*. Toronto: Edwin Mellen Press, 1999.

McAllister, J. W. "Truth and Beauty in Scientific Reason." *Synthese* 78 (1989): 25–37.

McCluskey, Stephen C. "Gregory of Tours, Monastic Timekeeping, and Early Christian Attitudes to Astronomy." *Isis* 81 (1990): 9–22.

McGuire, J. E. "Boyle's Conception of Nature." *Journal of the History of Ideas* 33 (1972): 523–42.

McKibben, William. *The End of Nature*. New York: Random House, 1989.

Macnaghten, Phil, and John Urry. *Contested Natures*. London: Sage Publications, 1998.

Mandelbrot, Benoit B. *The Fractal Geometry of Nature*. New York: W. H. Freeman, 1982.

Meilander, Gilbert. *The Taste for the Other: The Social and Ethical Thought of C. S. Lewis*. Grand Rapids, Mich.: Eerdmans, 1998.

Merchant, Carolyn. *The Death of Nature: Women, Ecology, and the Scientific Revolution.* New York: Harper & Row, 1980.

Midgley, Mary. *Evolution as a Religion: Strange Hopes and Stranger Fears.* Oxford and New York: Methuen, 1985.

———. *Science as Salvation: A Modern Myth and Its Meaning.* London: Routledge, 1992.

Miller, Perry. "The Romantic Dilemma in American Nationalism and the Concept of Nature." *Harvard Theological Review* 48 (1955): 239–54.

Modiano, Raimonda. *Coleridge and the Concept of Nature.* Tallahassee, Fla.: Florida State University Press, 1985.

Moltmann, Jürgen. *God in Creation: A New Theology of Creation and the Spirit of God.* Minneapolis: Fortress Press, 1990.

Mornet, Daniel. *Le Sentiment de la Nature en France de J.-J. Rousseau à Bernardin de Saint-Pierre: Essai sur les Rapports de la Littérature et des Moeurs.* Paris: Hachette, 1907.

Mosse, George L. *The Crisis of German Ideology: Intellectual Origins of the Third Reich.* New York: Howard Fertig, 1998.

———. "The Mystical Origins of National Socialism." *Journal of the History of Ideas* 22 (1961): 60–90.

Nash, James. *Loving Nature: Ecological Integrity and Christian Responsibility.* Nashville, Tenn.: Abingdon, 1991.

Newell, William Lloyd. *The Secular Magi: Marx, Freud, and Nietzsche on Religion.* New York: Pilgrim Press, 1986.

Odom, H. H. "The Estrangement of Celestial Mechanics and Religion." *Journal of the History of Ideas* 27 (1966): 533–58.

O'Hear, Anthony. *Beyond Evolution: Human Nature and the Limits of Evolutionary Explanation.* Oxford: Clarendon Press, 1997.

Ollman, Bertell. *Alienation: Marx's Conception of Man in Capitalist Society.* Cambridge: Cambridge University Press, 1977.

Pais, Abraham. *Niels Bohr's Times, in Physics, Philosophy and Polity.* Oxford: Clarendon Press, 1991.

Pannenberg, Wolfhart. *Toward a Theology of Nature: Essays on Science and Faith.* Philadelphia: Westminster/John Knox, 1993.

Penrose, Roger. "The Role of Aesthetics in Pure and Applied Mathematical Research." *Bulletin of the Institute of Mathematics and Its Applications* 10 (1974): 266–71.

Peterson, Ivars. *Newton's Clock: Chaos in the Solar System.* New York: W. H. Freeman, 1993.

Petruccioli, Sandro. *Atoms, Metaphors and Paradoxes: Niels Bohr and the Construction of a New Physics.* Cambridge: Cambridge University Press, 1993.

Pinnock, Clark H. *Flame of Love: A Theology of the Holy Spirit.* Downers Grove, Ill.: InterVarsity Press, 1996.

Polkinghorne, John. *Reason and Reality.* London: SPCK, 1991.

Porter, Carolyn. "Method and Metaphysics in Emerson's 'Nature.'" *Virginia Quarterly* 55 (1979): 517–30.

Post, Werner. *Kritik der Religion bei Karl Marx.* München: Kösel, 1969.

Postman, Neil. *Technopoly: The Surrender of Culture to Technology.* New York: Knopf, 1992.

Preus, Samuel J. *Explaining Religion: Criticism and Theory from Bodin to Freud*. New Haven, Conn.: Yale University Press, 1987.

Prigogine, Ilya, and Isabelle Stengers. *Order out of Chaos: Man's New Dialogue with Nature*. New York: Bantam Books, 1984.

Reid, Michael. "The Call of Nature." *Radical Philosophy* 64 (1993): 13–18.

Roe, Nicholas. *The Politics of Nature: Wordsworth and Some Contemporaries*. New York: St. Martin's Press, 1992.

Rorty, Richard. *Philosophy and the Mirror of Nature*. Princeton, N.J.: Princeton University Press, 1979.

Rosen, Joe. *Symmetry Discovered: Concepts and Applications in Nature and Science*. Cambridge and New York: Cambridge University Press, 1975.

Ruder, Cynthia Ann. *Making History for Stalin: The Story of the Belomor Canal*. Gainesville, Fla.: University Press of Florida, 1998.

Ruler, J. A. van. *The Crisis of Causality: Voetius and Descartes on God, Nature, and Change*. Leiden: E. J. Brill, 1995.

Ryle, Gilbert. *The Concept of Mind, Senior Series*. London: Hutchinson, 1949.

Schaeffer, Francis A. *Pollution and the Death of Man: The Christian View of Ecology*. Wheaton, Ill.: Tyndale House, 1970.

Schillebeeckx, Edward. *The Eucharist*. London: Sheed & Ward, 1968.

Schleip, Holger, ed. *Zurück zur Natur-Religion?* Freiburg: Hermann Bauer Verlag, 1986.

Schlesinger, George N. *The Intelligibility of Nature*. Aberdeen, Scotland: Aberdeen University Press, 1985.

Schmidt, Alfred. *The Concept of Nature in Marx*. London: New Left Books, 1971.

Schor, Juliet. *The Overspent American: Upscaling, Downshifting, and the New Consumer*. New York: Basic Books, 1998.

Schouls, Peter A. *Descartes and the Enlightenment*. Edinburgh: Edinburgh University Press, 1989.

Schreiner, Susan Elizabeth. *The Theater of His Glory: Nature and the Natural Order in the Thought of John Calvin*. Durham, N.C.: Labyrinth Press, 1991.

Schuffenhauer, Werner. *Feuerbach und der Junge Marx: Zur Entstehungsgeschichte der Marxistischen Weltanschauung*. Berlin: Deutscher Verlag der Wissenschaften VEB, 1972.

Schumacher, E. F. *A Guide for the Perplexed*. New York: Harper & Row, 1977.

Sheldrake, Philip. *Living between Worlds: Place and Journey in Celtic Spirituality*. London: Darton Longman & Todd, 1995.

Sircello, Guy. *A New Theory of Beauty*. Princeton, N.J.: Princeton University Press, 1975.

Smith, Huston. *Why Religion Matters: The Fate of the Human Spirit in an Age of Disbelief*. New York: HarperCollins, 2001.

Sobel, Dava. *Longitude: The True Story of a Lone Genius Who Solved the Greatest Scientific Problem of His Time*. New York: Penguin, 1996.

Soper, Kate. *What Is Nature? Culture, Politics and the Non-Human*. Oxford: Basil Blackwell, 1995.

Soulé, Michael E. "The Social Siege of Nature." In *Reinventing Nature: Responses to Post-*

modern Deconstruction, edited by Michael E. Soulé and Gary Lease, 137–70. Washington, D.C.: Island Press, 1995.

Stern, Fritz. *The Politics of Cultural Despair: A Study in the Rise of the Germanic Ideology*. Berkeley, Calif.: University of California Press, 1961.

Stewart, Ian. *Nature's Numbers: Discovering Order and Pattern in the Universe*. London: Weidenfeld & Nicolson, 1995.

Stout, Jeffrey. *The Flight from Authority: Religion, Morality and the Quest for Autonomy*. Notre Dame, Ind.: University of Notre Dame Press, 1981.

Talbot, John H. *The Nature of Aesthetic Experience in Wordsworth*. New York: Peter Lang, 1989.

Tatarkiewicz, Wladislaw. "The Great Theory of Beauty and Its Decline." *Journal of Aesthetics and Art Criticism* 31 (1972): 165–80.

Trousson, Raymond. *Le Thème de Prométhée dans le Littérature Européene*. Geneva: Droz, 1976.

van Bavel, Tarsicius. "The Creator and the Integrity of Creation in the Fathers of the Church." *Augustinian Studies* 21 (1990): 1–33.

Vogel, Steven. *Against Nature: The Concept of Nature in Critical Theory*. Albany, N.Y.: State University of New York Press, 1996.

Walsh, John, and Cynthia P. Schneider. *A Mirror of Nature: Dutch Paintings from the Collection of Mr. and Mrs. Edward William Carter*. 2nd ed. New York: Los Angeles County Museum of Art, 1992.

Weinberg, Steven. *Dreams of a Final Theory: The Search for the Fundamental Laws of Nature*. London: Hutchinson Radius, 1993.

———. *The First Three Minutes: A Modern View of the Origin of the Universe*. New York: Harper, 1993.

Weiner, Douglas R. *Models of Nature: Ecology, Conservation and Cultural Revolution in Soviet Russia*. Pittsburgh, Pa.: University of Pittsburgh Press, 2000.

Weisheipl, James A. "Aristotle's Concept of Nature: Avicenna and Aquinas." In *Approaches to Nature in the Middle Ages*, edited by Lawrence D. Roberts, 137–60. Binghamton, N.Y.: Center for Medieval and Early Renaissance Studies, 1982.

Westfall, Richard S. "The Scientific Revolution of the Seventeenth Century: A New World View." In *The Concept of Nature*, edited by John Torrance, 63–93. Oxford: Oxford University Press, 1992.

Wheen, Francis. *Karl Marx*. London: Fourth Estate, 1999.

Whitaker, Virgil K. *The Mirror up to Nature: The Technique of Shakespeare's Tragedies*. San Marino, Calif.: Huntington Library, 1965.

White, Andrew Dickson. *The Warfare of Science*. London: Henry S. King & Co., 1867.

White, Lynn. "The Historical Roots of Our Ecological Crisis." *Science* 155 (1967): 1203–7.

Whitehead, Alfred North. *Process and Reality: An Essay in Cosmology*. Cambridge: Cambridge University Press, 1929.

Whitney, Elspeth. "Lynn White, Ecotheology and History." *Environmental Ethics* 15 (1993): 151–69.

Woodbridge, Frederick J. E. *An Essay on Nature*. New York: Columbia University Press, 1940.

Worster, Donald. "Nature and the Disorder of History." In *Reinventing Nature: Responses to Postmodern Deconstruction*, edited by Michael E. Soulé and Gary Lease, 65–86. Washington, D.C.: Island Press, 1995.

Wutrich, Timothy R. *Prometheus and Faust: The Promethean Revolt in Drama from Classical Antiquity to Goethe*. Westport, Conn.: Greenwood Press, 1995.

Yoder, Joella G. *Unrolling Time: Christiaan Huygens and the Mathematization of Nature*. Cambridge: Cambridge University Press, 1988.

Young, Robert M. "Darwin's Metaphor: Does Nature Select?" *Monist* 55 (1971): 442–503.

———. *Darwin's Metaphor: Nature's Place in Victorian Culture*. Cambridge: Cambridge University Press, 1985.

Zee, A. *Fearful Symmetry: The Search for Beauty in Modern Physics*. New York: Macmillan, 1986.

Zimbardo, Rose A. *A Mirror to Nature: Transformations in Drama and Aesthetics, 1660–1732*. Lexington, Ky.: University Press of Kentucky, 1986.

Index